1970

his ʼ ʼl may be ke ʼ

The
Computer
Impact

Prentice-Hall
Series in Automatic Computation
George Forsythe, editor

ARBIB, *Theories of Abstract Automata*
BATES AND DOUGLAS, *Programming Language/One*
BAUMANN, FELICIANO, BAUER, AND SAMELSON, *Introduction to ALGOL*
BLUMENTHAL, *Management Information Systems*
BOBROW AND SCHWARTZ, EDITORS, *Computers and the Policy-Making Community: Applications to International Relations*
BOWLES, EDITOR, *Computers in Humanistic Research*
CRESS, DIRKSEN, AND GRAHAM, *FORTRAN IV with WATFOR*
DANIEL, *Theory and Methods for the Approximate Minimization of Functionals*
DESMONDE, *Computers and Their Uses, 2nd ed.*
EVANS, WALLACE, AND SUTHERLAND, *Simulation Using Digital Computers*
FIKE, *Computer Evaluation of Mathematical Functions*
FIKE, *PL/1 for Scientific Programmers*
FORSYTHE AND MOLER, *Computer Solution of Linear Algebraic Systems*
GAUTHIER AND PONTO, *Designing Systems Programs*
GOLDEN, *FORTRAN IV: Programming and Computing*
GOLDEN AND LEICHUS, *IBM 360: Programming and Computing*
GORDON, *System Simulation*
GREENSPAN, *Lectures on the Numerical Solution of Linear, Singular and Nonlinear Differential Equations*
GRISWOLD, POAGE, AND POLONSKY, *The SNOBOL4 Programming Language*
HARTMANIS AND STEARNS, *Algebraic Structure Theory of Sequential Machines*
HULL, *Introduction to Computing*
JOHNSON, *System Structure in Data, Programs, and Computers*
KIVIAT, VILLANUEVA, AND MARKOWITZ, *The SIMSCRIPT II Programming Language*
LOUDEN, *Programming the IBM 1130 and 1800*
MARTIN, *Design of Real-Time Computer Systems*
MARTIN, *Programming Real-Time Computer Systems*
MARTIN, *Telecommunications and the Computer*
MARTIN, *Teleprocessing Network Organization*
MARTIN AND NORMAN, *The Computerized Society*
MATHISON AND WALKER, *Computers and Telecommunications: Issue in Public Policy*
MCKEEMAN, HORNING, AND WORTMAN, *A Compiler Generator*
MINSKY, *Computation: Finite and Infinite Machines*
MOORE, *Interval Analysis*
PYLYSHYN, *Perspectives on the Computer Revolution*
PRITSKER AND KIVIAT, *Simulation with GASP II: A FORTRAN Based Simulation Language*
SAMMET, *Programming Languages: History and Fundamentals*
STERLING AND POLLACK, *Introduction to Statistical Data Processing*
TAVISS, *The Computer Impact*
TRAUB, *Iterative Methods for the Solution of Equations*
VARGA, *Matrix Iterative Analysis*
VAZSONYI, *Problem Solving by Digital Computers with PL/1 Programming*
WILKINSON, *Rounding Errors in Algebraic Processes*

The
Computer
Impact

EDITED BY
IRENE TAVISS

Program on Technology and Society
Harvard University

PRENTICE-HALL, INC., Englewood Cliffs, New Jersey

Printed in the United States of America

C — 13-165969-3
P — 13-165936-7

Library of Congress Catalog Card No.: 74-128776

Current printing (last digit):
10 9 8 7 6 5 4 3 2 1

PRENTICE-HALL INTERNATIONAL, INC., *London*
PRENTICE-HALL OF AUSTRALIA, PTY. LTD., *Sydney*
PRENTICE-HALL OF CANADA, LTD., *Toronto*
PRENTICE-HALL OF INDIA PRIVATE LIMITED, *New Delhi*
PRENTICE-HALL OF JAPAN, INC., *Tokyo*

To My Parents

Preface

To take on the task of setting forth the social impacts of computer technology requires either extreme gall or much naiveté. The subject is both so enormous and complex and so changing that no definitive statement could possibly be made at this time. Disclaiming both gall and naiveté, I offer this reader modestly, in the hope that it will serve to clarify some issues, raise questions about others, and explain some of the many dimensions of this complex and highly important subject.

My thinking on the more general questions of the relationship between technology and society has been influenced greatly by my colleagues at the Harvard University Program on Technology and Society: Emmanuel G. Mesthene, Director and Juergen Schmandt, Associate Director. I am indebted also to two associates and friends of the Program who have worked more closely with problems of computer technology: Anthony G. Oettinger, Professor of Linguistics and Applied Mathematics at Harvard University and Alan Westin, Professor of Public Law and Government at Columbia University.

I.T.

Contents

Part IV   The Polity

Part V The Culture

The
Computer
Impact

I

GENERAL INTRODUCTION

General Introduction

If one looks at the most prominent reactions to computers in our society—from the "do not fold, staple, or mutilate" protest signs to the excitement generated by computerized space capsules—the most appropriate title for a discussion of the social implications of computer technology might appear to be *Computers: Curse or Blessing?* The computer has excited the public imagination and has generated both great fears and great hopes. It has become a symbol for all that is good and all that is evil in modern society.

REACTIONS TO COMPUTER TECHNOLOGY

Why does the computer evoke such reaction? Part of the answer lies in the aura of all-pervasiveness which surrounds it. As a tool which can store, analyze, manipulate, and present data in a variety of ways, it has applications in almost all spheres of human activity.[1] Its efficiency in helping man to cope with the "information explosion" and to impose some order on complexity makes it a highly attractive device for a large number of organizations. Business firms, schools, government agencies, churches, hospitals, research laboratories, and law firms all find uses for computers. The individual, in turn, finds his checks, his income tax returns, his applications for jobs or credit cards, his magazine subscriptions, and his political opinions processed by computers. He also has a vague awareness that in the future his children, or perhaps his grandchildren, will be educated by a computer, that his informational needs will be serviced by computers, and that his financial transactions will be taken care of by an electronic cash and credit system. With all of this as background, he may fear that the essentials of his personal life will be stored in the memory bank of some computer and be used for purposes beyond his control. Or, if he is of an optimistic bent, he may view computers as machines which will help political decision-makers to program the ultimate in social wisdom and justice.

[1] A radio advertisement which celebrates the efficiency of the "new computerized railway" portrays a teacher asking her class what computers do. One child answers enthusiastically that "computers help do everything."

A second reason for the high degree of emotional reaction to computers is the challenge or competition which they seem to present to man. While everyone who is even slightly familiar with the operation of computers recognizes that their effectiveness depends upon the intelligence and creativity of the men who program and use them, the computer specialist's classic "GIGO"—"Garbage In, Garbage Out"—is all too often replaced by another "GIGO"—"Garbage In, Gospel Out." If not deified, the computer is at least seen as human or superhuman. That it is often viewed as a competitor to man is seen, for example, in the great delight that is evoked by computer errors. No other machine failures elicit the same reaction. When a computer "makes a mistake," man is reassured that this machine is as fallible as he is. Yet the same powers of the computer which generate hostility in some men evoke hope in others, who view the computer as an extension of man's reasoning powers and the man-computer relationship as a cooperative and beneficial one.

Those who choose to view the computer with hostility often see it as a "dehumanizing" agent. From this perspective it is not principally a useful tool or an important labor-saving device, but a machine which imposes its own logic on men. Exactly what makes the computer "dehumanizing" is usually not stated, but the complaint appears to be based in part on the more general grievance that machines exert some influence on man's behavior. Some people have to work nights to keep expensive computers in operation; others have to work more carefully and accurately than they were used to; still others are required to collect and process information in a certain way. That is, to derive maximum benefit from computers, men must in some ways adapt to them. But this is true of most tools and machines that men build and use.

Computers may also be seen as "dehumanizing" because they produce conclusions that are unexpected or do not confirm the assumptions of the men who fed the data into them. The whole enterprise of science may be seen as "dehumanizing" in this sense, since the results of scientific research often turn up findings that are contrary to man's expectations or traditional beliefs. It may be that computers have become a convenient target for those who are unhappy with the role that both science and sophisticated machinery play in our society.

There is one final sense in which the computer may be seen as "dehumanizing." Peculiar to the computer itself, this sense provokes the protest signs that read "do not fold, staple, or mutilate." It is the reason that the "IBM card" has become a symbol of "dehumanization" or "depersonalization." What provokes protest here is the fact that information about persons can be classified and coded into preset categories for computer processing. The real target of this complaint should be the social complexity that results from large populations in advanced technological societies. With or without computers, the need to process large amounts of information efficiently would still be present. Indeed, without computers the categories might be cruder and the rigidities of large organizations much greater. Ironically, it may be that the computer allows for more "individualized" treatment than would otherwise be feasible.

While the anti-computer spirit might be somewhat more prominent in the mass media today than the awe-and-wonder stance, some equally exaggerated conceptions prevail on the other side. Chief among these is the notion that computers can "do everything" and that the use of computers will go a long way towards solving our basic problems in education, medical care, organizational efficiency, and political decision-making. The extreme optimists help to feed the anti-computer spirit by making computers appear to be precisely the kind of autonomous superbrains that those who are hostile to them so often fear. Moreover, by proclaiming that the use of computers will bring rapid and beneficial social changes, the optimists generate unrealistic expectations and kindle activities that are wasteful because they are undertaken too hastily and without adequate recognition of the difficulties of effecting social change.

The polarization of views concerning the impact of computers has led to exaggerations, distortions, and obfuscations of the complex issues involved. The notion, for example, that computers are "throwing people out of work" has turned out to be an ungrounded and overly simple fear. It is the rate of economic growth, rather than technological change *per se,* which determines the level of unemployment. While specific technological changes might be responsible for the displacement of particular categories of workers, the ability to find other employment depends upon economic policies which sustain an adequate rate of economic growth. Predictions, which were rampant in the late 1950's and early 1960's, that computer technology would soon effect revolutionary changes in the numbers of people employed and in the length of the work week have not been realized. Predictions of major and rapid changes in such fields as educational technology and management techniques have been similarly exaggerated.

It is not the technology itself, but the society, or groups within it, which by their policies and allocation of resources, determine the speed and nature of particular social changes. This is not to deny that the development of new tools may in some instances have unforeseen consequences. The adaptations and changes which men make in order to derive benefit from the use of a new tool often result in further changes. Just as the use of automobiles triggered a variety of social changes which were not intended or foreseen when automobiles were first introduced, so might the use of computers. One example which has often been cited is the manner in which computers have helped to change the nature of the banking business. Modern banks provide a large number of services that were previously not possible and not even considered as functions that banks might perform (e.g., handling a customer's accounts and billing).

What then can be said about the impact of computer technology on society? Two general points stand out. Powerful tools always have powerful effects; and social changes occur to take advantage of the benefits that the tools might provide. At the same time, however, social factors mediate between a tool and its effects.

COMPUTER IMPACTS

The impact of computers cannot be appreciated without some understanding of the society into which they are being introduced. The development and application of computer technology would not have been likely if ours were not a society that is knowledge-oriented, concerned with innovation and planning, and informed by a scientific ethos. At the same time, the use of computers generates new opportunities and new problems to which the social structure must respond. Since the changes set in motion by computers are very current, it is, in many cases, too early to discern their direction.

The Economy

The modern business corporation, like the society of which it is a part, is larger and more complex than its counterpart of an earlier day. As the number and size of its specialized parts and the extent of its interaction with other large organizations (unions, governments, universities) have increased, more sophisticated coordinating and planning mechanisms have become necessary. Since corporate growth has become at least as important to modern businesses as the profit motive has traditionally been, more resources have come to be devoted to research and development and to innovation.

These changes have resulted in a reversal of the relative importance of capital and labor. While capital may have been most important in the past, today labor—talented personnel—has become crucial to the success of the corporation; and knowledge has become a critical factor in this success. As Daniel Bell has noted, our society may be characterized as a "knowledge society" in two senses: "first, the sources of innovation are increasingly derivative from research and development...; and second, the 'weight' of the society—measured by a larger proportion of Gross National Product and a larger share of employment—is increasingly in the knowledge field."[2]

Not only is the management of the modern business corporation devolving increasingly upon technically trained and highly specialized personnel,[3]

[2] Daniel Bell, "The Measurement of Knowledge and Technology," in Eleanor Bernert Sheldon and Wilbert E. Moore, eds., *Indicators of Social Change* (New York: Russell Sage Foundation, 1968), p. 198.
[3] See Jay M. Gould, *The Technical Elite* (New York: Augustus M. Kelley, 1966).

but throughout the occupational structure, education and training have assumed a new importance. This has occurred as a result of the progressive elimination of the less skilled occupations and the growth of professional, technical, and white-collar work.

While the increased importance of educational attainments for entry into higher level positions is not in dispute, the importance of education as a prerequisite for employment is a matter of controversy. In a society with sufficient wealth to allow large numbers of people to postpone entry into the labor force so that they may receive more education, employers often have a sizeable pool of educated manpower at their disposal. It is true that as technological advances have diminished the need for unskilled manual labor, many of the newer jobs have tended to require greater education and training. But the educational requirements for a given job are often exaggerated. Given the value placed upon education in our society and the ease with which educational credentials may serve as a screening device, employers will often prefer to hire the more educated, even though the intrinsic requirements of the job do not call for such education.

Much of the growth in professional, technical, and white-collar employment has resulted from the growth of the service sector. Technological advance in the production and processing of goods has allowed the growing demand for goods to be met by an increasingly smaller proportion of the total work force. At the same time, as income has risen, the demand for services has increased more rapidly than the demand for goods. While the percentage of gross national product represented by the service sector remains considerably lower than that devoted to the goods sector, over half of the labor force has been employed in the services since the 1950's.

It would appear that the upgrading and specialization of the labor force, the trend towards managerial planning and coordination, and the growth of the service sector are all enhanced by computer technology. Yet caveats and qualifications are in order in each case.

The introduction of computers into factories appears to eliminate many of the routine and low-level jobs. But whether the average skill level of the blue-collar worker is raised—because lower-level jobs are progressively eliminated—or lowered—because the machines also take over some of the work which had been done by skilled workers—is a matter of some dispute. In some instances, for example, when an operator requires less skill because of the introduction of automated machinery, he may be made responsible for a larger portion of the production sequence, and this assignment may require a knowledge of additional machines.[4] The consensus appears to be

4 James R. Bright, "The Relationship of Increasing Automation and Skill Requirements," in National Commission on Technology, Automation, and Economic Progress, *The Employment Impact of Technological Change,* Appendix Volume II to *Technology and the American Economy* (Washington, D.C.: Government Printing Office, 1966), pp. 203–221.

that automation raises the average skill level in some respects and lowers it in others, so that there is no substantial net change.[5]

In the case of office workers, it appears that the introduction of computers results in an upgrading of skills.[6] Bottom level clerical jobs are eliminated and a new level of office workers develops around the computer—technicians, programmers, systems analysts.

However, in order to make the most efficient use of computers, some clerical workers are now having to work on second and third shifts—a phenomenon clearly associated with lower status, blue-collar jobs; and working with computers may require more manual labor than is associated with traditional clerical work. At the same time, the blue-collar workers in computerized industries acquire greater responsibility and are called upon to exercise skills which are nonmanual in nature. Hence, the distinction between blue- and white-collar employment becomes blurred. While the salary differential between the two groups has long been a small one, and in many cases blue-collar workers have earned more than lower level white-collar workers, the status differential had not previously been challenged. In computerized industries, this differential is beginning to disappear. For, "where the ratio of capital to labour costs is high, maximum utilisation of plant becomes extremely important. It becomes even more necessary to avoid breakdowns and to have a reliable labour force. Granting such 'white-collar' conditions as an annual salary [and] improved fringe benefits is then seen partly as the price to pay for dependability. . . . It is also in such industries that technology has had, and is likely to have most effect upon the content of the job. The dividing line between the skilled plant operator who is 'blue-collar,' and the technologist who is 'white-collar' becomes very narrow indeed."[7]

As blue- and white-collar jobs begin to converge, a new stratum of personnel is emerging around the computer. The computer professionals often pose a threat to the older managerial personnel. Their training and behavior are different from those of the managers; and they usually maintain a dual loyalty—to their profession and to the organization which employs them. "The resistance and status anxiety" which managers exhibit

---

[5] See *ibid.* and Morris A. Horowitz and Irwin L. Herrnstadt, "Changes in the Skill Requirements of Occupations in Selected Industries," *ibid.,* pp. 223–287; Edwin Mansfield, *The Economics of Technological Change* (New York: W.W. Norton and Co., 1968), pp. 134–161; Joseph N. Froomkin, "Automation," in *International Encyclopedia of the Social Sciences* (New York: The Free Press, 1968), Vol. 1, pp. 480–489.

[6] See Roy B. Helfgott, "EDP and the Office Work Force," *Industrial and Labor Relations Review,* 19 (July 1966), pp. 503–516 and Kenneth F. Walker, "Personal and Social Planning at the Plant Level," Document D 11–68 of the Third International Conference on Rationalization, Automation, and Technological Change, sponsored by the Metalworkers' Industrial Union of the Federal Republic of Germany (Oberhausen, Germany, 1968), 26 pp.

[7] Dorothy Wedderburn, "Are White-Collar and Blue-Collar Jobs Converging?," Document P 12–68 of the Third International Conference on Rationalization, Automation, and Technological Change, *op. cit.,* p. 20.

as computers are introduced "arises out of the conflict between older, organisation-oriented and often tradition-bound managers and a profession-oriented, machine-minded, often radical and innovating 'computer elite.' "[8]

It has been suggested that these tensions might be alleviated if managers received more technical training. The advent of office automation has raised much discussion about "how specialised managers ought to be and how far 'professionalisation' of management should be allowed, or encouraged. . . . The prevalent view seems to be that technical background is not likely to become a prerequisite for top managers. While some students of the subject incline to the view that top managers must understand mathematical concepts, it is not generally thought that they themselves need particular skills in their application. It is coming to be generally accepted that top executives need the best possible general educational background."[9] Yet it has been found that "new men are beginning to emerge at the very highest levels of command of America's leading corporations, men who speak the language not only of science and engineering but of business as well."[10] This increasing prominence of technical managers has been attributed to the fact that "key managerial decisions today rest increasingly on technical and scientific premises that impinge upon and frequently override financial, marketing, and other business considerations."[11]

The extent to which managerial decisions are affected by the use of computers is also a matter of debate. A careful examination of the ways computers are used in more than a dozen large companies revealed that computers had no direct impact on top management. The computers were used by middle management in giving advice to top level management.[12] Moreover, the computers were of only limited assistance in decision-making. The decision-making process may be seen as consisting of the following steps: identify a problem, analyze it, set forth the alternatives, evaluate the alternatives, and select a course of action. Viewed from this perspective, it was found that computers were used more frequently in analyzing the situation than in identifying the problem and more frequently in identifying the problem than in evaluating the possible courses of action; "no cases were observed where the computer was used in defining the possible courses of action."[13]

Aside from the fact that some decisions simply do not lend themselves to computer analysis, the obstacles to taking fuller advantage of the potentials of computers as aids in decision-making include inadequacies in computer hardware and software; insufficient understanding of currently unquantifiable information; a lack of knowledge, or a defensive attitude

[8] H.A. Rhee, *Office Automation in Social Perspective* (Oxford: Basil Blackwell, 1968), p. 76.
[9] *Ibid.*, pp. 130–131.
[10] Gould, *op. cit.*, p. 77.
[11] *Ibid.*, p. 84.
[12] Rodney H. Brady, "Computers in top-level decision making," *Harvard Business Review*, 45 (July–August 1967), pp. 67–76.
[13] *Ibid.*, p. 70.

against, the ways in which computer information can be used in making decisions; and sometimes an unwillingness "to incur the expense and risk of pioneering and testing new areas of computer application."[14]

But even if the extent of management's reliance on computers has often been exaggerated, the organizational consequences of introducing computers are often quite substantial. While computers help to digest and simplify the vast amounts of information with which all modern organizations must deal, they also generate massive outputs that must be read, digested, and analyzed. Determining whose function this should be, and how the various computer personnel should relate to other members of the organization, often poses extremely difficult problems. Also, adjustment must be made to the changed pace of work and the greater need for accuracy that are imposed by computer technology. The resolution of these problems—the ways in which an organization reacts to the new situation created by computer technology—depends, in large part, upon the nature of the organization prior to the introduction of computers.

One major issue that has been a source of much controversy is the extent to which the introduction of computers leads to greater organizational centralization. While centralization often seems to result from office automation, this is not an inevitable outcome. For organizational structure—the division of labor, levels of hierarchy, and channels of communication—intervenes between automation and its effects. Thus, it is only in traditional organizations which do not allow for "horizontal" channels of communication that disputes between the computer experts and the "line" managers will tend to be resolved by the head of the organization, thereby leading to a high degree of centralization of control. When "an organization can dispense with the concept of a rigid chain of command and allow horizontal interchange, then automation will have little impact on its members' ability to co-operate with one another," and such centralizing tendencies will not emerge.[15]

Clearly, computer technology "leads to more relevant and precise information being available—and much more quickly—to central management and . . . as a result, central management must exercise more analytical judgement. This would seem to imply more centralised policy control. . . . But within the framework of central control of general policy there remains much scope for decentralised operating authority—*i.e.,* a better factual basis for local decision-making within the limits of centralised policy control. . . . [Moreover,] smaller computers have become cheaper and more readily available. They have tended to be placed in decentralised locations. Small computers are useful (i) for processing data which are highly local in character and may have little relevance for upper management decision-making, and (ii) for preliminary treatment of data, before transmission to the central system."[16]

14 *Ibid.,* p. 76.
15 Marshall W. Meyer, "Automation and Bureaucratic Structure," *American Journal of Sociology,* 74 (November 1968), p. 264.
16 Rhee, *op. cit.,* p. 109.

These countervailing tendencies within organizations are also found in the economy at large. While the overall and long-term trend seems to be towards greater centralization, movement in this direction is by no means inevitable.

The one economic trend which seems to be in least contention is the growth of employment in the service sector. The only debate here concerns the rapidity and extent of this growth. While most observers have been impressed by the fact that by the 1950's over half of the United States labor force was employed in the services, it has been argued that the shift has been occurring slowly and will continue to progress gradually.[17]

It is likely that as the demand for services continues to increase, more attention will be devoted to the application of technology in these areas. Computers are being introduced into such diverse areas as transportation and communication, banking and finance, education, and medical care. As their potential begins to be utilized, new services tend to arise. Not only are billing and accounting procedures computerized, but airline and hotel reservations have become computerized, and such diverse human needs as finding employment and finding a mate have—in a limited way, to be sure—become computerized services. As a result of such new activities, new jobs are being created as older jobs are displaced.

Some anticipated employment increases have not occurred, however, because of the difficulties and constraints involved in restructuring some traditional services. It is probably safe to say that the major welfare services have not yet been affected significantly by computers. If schools and hospitals use computers for various record-keeping purposes, the nature of education and of medical care have not been altered. The promise and the dangers of computers in the classroom and hospital are still matters of exploration and debate. Wide-scale and meaningful use of computers in these areas would require radical transformation of the organizations and professions in question. These do not occur easily.

The Polity

Some radical changes in organization would also be required if computerized information systems were to be used as efficiently and effectively in the political decision-making process as their advocates have hoped. While the adoption of such systems makes sense in the context of a political structure which is increasingly oriented towards planning and the most effective use of knowledge, some of the thorniest issues of contemporary political life arise in this connection.

Like the modern business corporation, the modern government has responded to greater complexity by developing more sophisticated mechanisms for coordination and planning. Governmental responsibility has in-

---

[17] See Leon Greenberg, "Is A Service Society Developing?," paper presented at the Third International Conference on Rationalization, Automation, and Technological Change, *op. cit.,* 42 pp.

creased as a result of affluence and the development of large-scale technologies. In an affluent society, social welfare expectations are heightened, so that many goods and services which were previously considered matters of privilege are now seen as rights. The government thus acquires the responsibility for assuring that such amenities as education and medical care are provided to all members of the society. The need to control such effects of technology as pollution and radiation has likewise led to the increased involvement of government as regulatory and control agent.

The government has become not only a user, but also a major purchaser of scientific and technical knowledge. The accumulation and use of knowledge have helped to bring pressures for greater social planning because both the new intellectual tools that are being developed (e.g., systems analysis, computer simulations, and planning-programming-and-budgeting systems) and the idea of deliberate creation of knowledge to help solve (or prevent) social problems have begun to take hold.

Decision-makers at all levels of government have become more conscious of the need for studying the consequences of proposed actions. Advisory groups have been established to help plan and assess government policies. It is in this context that computers have been seen as potentially useful tools. But the attempt to take advantage of this potential poses many problems.

To establish an information system for use in administrative or decision-making processes requires the collection of accurate information. If an information system is to be set up within one government agency, this first step may pose few problems. If, however, the pooling of information from a variety of agencies or departments is required, some difficulties may ensue. Such sharing of information files would not only reduce costly duplications; it would also provide a more comprehensive data base. But the pooling of information across established jurisdictional boundaries arouses fears that the authority of the agencies in question will be diminished. In order to integrate information from diverse sources, new consolidated organizations may be needed. The problem is that "parochialism inhibits integration across various levels of government, especially where organizational changes are necessary. . . . However, when problems become severe, political subdivisions will accept centralization: *e.g.,* metropolitan water districts, transit districts, and policy information networks."[18]

There are thus two first-order problems in establishing information systems: the vested interests and resistances to organizational change and the question of what new political structures might be needed. If the current organization of government results in gross overlaps, inefficiencies, and inadequate handling of problems, it is not clear what new mode of organization might remedy these problems while maintaining a federalist system of government. Some consolidation of authority at the local level seems appropriate. The United States today has 50 states, "3,043 counties, 17,144 towns and townships, 17,997 municipalities, 34,678 school districts, and

[18] See the essay by Paul Armer reprinted on pp. 123–129.

18,323 special districts. The number of separate agencies among these 91,236 areas apparently has never been recorded."[19]

Yet there is much concern about how far consolidation or centralization should proceed. Some observers believe that the federalist system might no longer be an appropriate way of organizing government. Conversely, it has also been argued that greater efforts must be made to protect and strengthen the authority of the states.

Centralization of information itself, even in the absence of major consolidations of authority, has raised some difficult problems. The proposal set forth in 1966 by the Bureau of the Budget to establish a National Data Center became the subject of violent controversy. Opponents feared that such a system would endanger individual privacy. Cries of "1984" were heard through the halls of Congress as the bill was being debated. The bill's proponents argued that both legal and technical safeguards could be built into such a system to insure against invasions of privacy. Although the intent of the proposal was to pool the information that exists in various government agencies and not to establish individual dossiers, it was feared by many that information about individuals could be acquired too easily under such a centralized system. Undoubtedly, the bill's opponents exaggerated the dangers of a national data bank, while its supporters minimized them. A modified version of the original proposal is still pending.

Fears about invasion of privacy are often part of a larger anxiety concerning the way in which decision-makers will use the information at their disposal. In an already highly complex society in which the range of what citizens can know and control has diminished, the sophisticated techniques of decision-making that accompany computerized information systems are seen as a further menace to the rights of the citizen. But visions of political takeover by the computer experts and allied scientists ignore what Alan Westin has called the "filtration process of American values, processes, and institutions."[20] That is, the changes in the political process which may result from computerized information systems will be determined in part by the nature of our political structure and values.

As has been noted, organizational resistance to change is often a very formidable factor militating against rapid change in response to a new technology. In addition to this barrier to the introduction of information systems, there are aspects of the rationality or ethic of American political life which run counter to the rationality of such techniques as systems analysis and planning-programming-and-budgeting systems. Systems analysis begins with some clearly defined objectives. A long-range goal is chosen and careful calculation and measurements are then made to determine what the programs established to implement this goal are actually achieving. The rationality of politics, on the other hand, often dictates a focus on more short-range goals, the "appearance" of effort, and a thin spreading

---

[19] *Ibid.*
[20] See the essay by Alan Westin reprinted on pp. 130–144.

of resources over many programs in order to gain support from diverse groups. Moreover, because there is a tendency on the part of government agencies to withhold information that puts them in an unfavorable light and to practice a certain fuzziness, if not distortion, in order to maintain their competitive positions, there are problems involved in the initial collection of information.[21] Systems analysis cannot be fruitful if the quality of its data base is poor. The experience thus far with the use of planning-programming-and-budgeting in the Department of Defense has led most observers to agree that despite the computers and the staff of analysts, little substantive change has occurred in the operation of the Department.

The promise of systems analysis and computerized information systems lies in their ability to make decision processes more rational. American society places a high value upon what has been called "secular rationality."[22] Yet fears abound that the limitations of the rationality involved in systems analysis will not be recognized. There is a suspicion that those elements of a decision which cannot be readily quantified will not be given due attention. While this is certainly a phenomenon to be guarded against, it should be noted that the systems analysts do not usually set the goals. Despite the exaggerated claims of some of its practitioners and advocates, systems analysis cannot handle irreconcilable objectives. The setting of goals is a matter to be handled by the political process. Thus, the fear that "human values" will be ignored in the new "rationalized" decision-making is somewhat exaggerated, since the problem of values is usually most salient in the choice of goals.

A second fear is that the improved rationality which might result from the new decision-making techniques would be at the expense of "democracy." It is felt that more and more decisions will be placed in the hands of the experts and the entire process will be far removed from the control of the people. In the face of "information overload" and the complexities involved in many political decisions, the question of what constitutes the proper balance between expert and popular authority is a most difficult one.

As Martin Shubik has put it: "Perhaps the eighteenth and nineteenth centuries will go down as the brief interlude in which the growth of communications and knowledge relative to the size of population, speed of social and political change, and size of the total body of knowledge encouraged individualism and independence. By its very success, this brought about the tremendous need for and growth of knowledge reflected in the research monasteries, colleges of specialists, and cloisters of experts of the twentieth century's corporate society."[23] He goes on to argue that although what we mean by "democracy" might change, computers will be a help rather than a hindrance in preserving democracy. "If we wish to preserve

21 James Schlesinger, "Systems Analysis and the Political Process," *Journal of Law and Economics,* 11 (October 1968), pp. 281–298.
22 Robin M. Williams, Jr., "Individual and Group Values," *Annals of the American Academy of Political and Social Science,* 371 (May 1967), pp. 20–37.
23 Martin Shubik, "Information, Rationality, and Free Choice in a Future Democratic Society," *Daedalus,* 96 (Summer 1967), pp. 772–773.

even modified democratic values in a multi-billion-person society, then the computer, mass data processing, and communications are absolute necessities. It must be stressed...that they are necessary, but not sufficient. Using an analogy from the ballet, as the set becomes more complex and the dancers more numerous, the choreography required to maintain a given level of coordination becomes far more refined and difficult. The computer and modern data processing provide the refinement—the means to treat individuals as individuals rather than as parts of a large aggregate."[24]

The crucial issue of the effects of the new technology on democratic government hinges in large part on the questions of who will have access to the data and what kinds of controls and safeguards will be established. These questions are important not only for the preservation of privacy, but also for the maintenance of a society in which all power does not reside in the government. Alan Westin has speculated that in the future, power may rest upon access to the best information. Hence, the right to information becomes extremely significant.

If the use of computers to store and analyze large quantities of information quickly may be used to enhance the importance of experts at the expense of citizen participation, the possibility also exists that computer technology could be used to establish a system of direct democracy. The American concept of democracy, reflected in mythologies of "New England town democracy," rests on the assumption that each citizen should ideally have a direct voice in political decision-making. Since this is not feasible, elected officials serve to represent his interests and opinions. However, as sophisticated computer technology increasingly makes possible the rapid and efficient collection of opinions, the possibility of instituting direct democracy is raised. A system of "instant voting" by the whole electorate on any issue presented to it could eventually be technically possible. Its political feasibility and its wisdom are other matters.

Short of the introduction of computerized "instant voting," the rapid collection and diffusion of public opinion data is already having some effects on our political system. It has also raised some questions about the meaning of leadership in a democratic society. The ideal of a democratic society holds that the leadership is to be responsive to the public will. Hence, the "populists" argue that government leaders should act in accordance with the results of public opinion polls. The "elitists," on the other hand, maintain that leadership consists in being "above" the pressures of public opinion. The populist/elitist divergence has long been a source of strain in the American political system, but the presence of technology to measure the "public will" has aggravated this tension.

The effects of public opinion polls on the practice of leadership are difficult to determine. There is no reliable evidence to indicate that leaders follow poll results blindly against their own better judgment. "Legislators are not particularly responsive to public opinion, chiefly because they become entrapped in the power struggle and archaic rules within their own estab-

[24] *Ibid.*, p. 777.

lishment."[25] But poll data may serve as an important political instrument. It has been suggested that the polls may help to thwart the power of pressure groups. On the birth control issue, for example, Catholic pressure groups would have been able to claim that all Catholics oppose the distribution of birth control information were it not for the fact that poll data show this to be false.[26]

The polls have also been of some importance in the electoral process. While some critics have alleged that the existence of better information on public attitudes and preferences allows for greater manipulation of the people, it could equally well be argued that the "interference" of the polls has helped to thwart "machine politicians." Moreover, the protection of the people against such "manipulation" lies, as it always has, in the hands of the people. It is the citizen's responsibility to be concerned about public issues, to keep himself well informed, and to avoid falling into stereotypic or simplistic patterns of response.

The Culture

The great difficulty of keeping "well informed" today results from the sheer amount of information, the number and speed of new discoveries, the sophistication and intricacies of highly specialized fields of knowledge, and the high degree of relatedness between events in different societies and different sectors of the same society. In a culture whose orientation appears to be increasingly cognitive and scientific, a high value is placed upon both the generation of new knowledge and the efficient utilization of existing knowledge.

The enhanced importance of scientific and theoretical knowledge has resulted from the need to understand, support, and control a complex and changing society. "A changing society must put a relatively strong accent on knowledge in order to offset the unfamiliarity and uncertainty that change implies. Traditional ways (beliefs, institutions, procedures, attitudes) may be adequate for dealing with the existent and known. But new technology can be generated and assimilated only if there is technical knowledge about its operation and capabilities, and economic, sociological, and political knowledge about the society into which it will be introduced."[27]

Not only is scientific knowledge increasingly relied upon for developing technologies and controlling their social effects, but the scientific mentality has become an important part of the ethos of modern society. A cultural climate has developed in which more and more assertions and beliefs are being subjected to scientific testing. And as more is learned about the consequences of various actions, "the positive evaluation of *cognitive* criteria

25 George Gallup, "Polls and the Political Process—Past, Present, and Future," *Public Opinion Quarterly*, 96 (Winter 1965–1966), p. 547.
26 *Ibid.*
27 Emmanuel G. Mesthene, "How Technology Will Shape the Future," *Science*, 161 (July 12, 1968), p. 138.

for judging both individual conduct and collective policy and practice" increases.[28]

As a tool for collating and analyzing information, the computer is an aid in the process of using knowledge. Computers help to reduce some of the complexities involved in handling large amounts of data, to render the data "more manageable." At the same time, however, they allow for more complex analyses and manipulations and foster newer, more complex modes of thinking. A "rapport with the computer, which will make it possible for the creative person to think in terms of many, many variable, probabilistic, and dynamic relationships is bound to produce multi-variable, probabilistic, and dynamic models of the world that couldn't be invented or evaluated otherwise."[29]

While such models may be of great assistance in dealing with complex problems, their existence may produce a new from of illiteracy. For "in all areas where logical model-building can enhance our understanding of social and material reality, we can expect the computer to enable men to create descriptions of that reality that will be essentially incomprehensible to those who are not part of the world where reality is mediated by the computer."[30] Hence, the need to wipe out "computer illiteracy" might become a major educational problem.

Some high schools and colleges have already instituted programs designed to familiarize their students with the way computers operate. But considerably less public attention has been given to this aspect of "computers and education" than to the use of computers in the schools for instructional purposes.

If one ignores the extremists on both sides—those who view computers as "the solution" to "the education problem" and those who deplore the "further dehumanization" to be produced by "substituting machines for live teachers"—the dispute about the role of computers in the classroom seems to hinge on questions regarding how much of the educational curriculum can profit from some form of computerization and how quickly we are progressing towards the meaningful use of computers in the schools. Despite the existence of some few showcase schools scattered across the country, and despite the lavish publicity they receive, the use of computers in the schools has thus far been very limited. Four basic factors appear to be responsible for this: the inadequacies of our knowledge about the learning process, the inadequacies of the technology, the financial constraints, and the extreme difficulty of effecting organizational change within most school systems.[31]

The ultimate promise held out by the advocates of computers in the classroom is that of "individualized instruction." In this model, students

---

[28] Williams, *op. cit.*, p. 30.
[29] Donald N. Michael, *The Unprepared Society: Planning for a Precarious Future* (New York: Basic Books, Inc., 1968), p. 49.
[30] *Ibid.*, p. 50.
[31] See Anthony G. Oettinger, *Run, Computer, Run: The Mythology of Educational Innovation* (Cambridge, Mass.: Harvard University Press, 1969).

would be able to learn at their own pace and teachers would serve to guide individual students and monitor their progress, and to lead various types of small group discussions and activities. But this "best of all possible worlds" remains a futuristic one. The coordinating problems in such a system would be enormous, due to the great variations among students in rate of learning. Moreover, the computer programs that would have to be written to allow for true individuality of responses are beyond our present capabilities. To this extent, there is some truth in the arguments of those who fear that standardization or homogenization would result from computer-aided instruction.

However, a close examination of current educational practices would probably reveal precisely the same kind of standardization. With much effort and careful research, the development and use of computers for instructional purposes might prove to be superior to our present system. There is much room for speculation about the changes which might occur in our educational system, because there is much room for improvement. Despite the many and profound social changes which have occurred since our system of public education was established, the way in which we educate our children has undergone very little change. In the face of rapid knowledge obsolescence, the changing nature of jobs, and the possibility that most persons in the future will have several different careers during the course of a lifetime, it is likely that real changes in our educational system will be forthcoming. Indeed, the very notion of a "school" might eventually be displaced in favor of some process that would integrate the educational, occupational, social, and personal development aspects of schooling. If and when such changes occur, computers might play a significant role.

Even in the absence of such changes in the formal educational process, much discussion is under way concerning the potential uses of computers for informal education and information dissemination. Again, there is a futuristic quality about this discussion since many hurdles—in the form of technical difficulties, and financial and organizational constraints—must first be overcome. But the vision includes such things as time-sharing systems in which large networks of information could be consulted by individuals in much the same way as they now use libraries. Computerized "newspapers" might be available so that individuals could ask for and receive in-depth background information about any particular story in the news which interests them. Similarly, computer tapes containing movies, plays, documentaries, and an information file might be tied to television sets so that the viewer could question the system to select what he wishes to see and then have the item run on his screen.[32]

In addition to the dissemination of information and knowledge, computers also help to generate knowledge. There is little doubt that research and scholarship have profited greatly from these powerful tools. If at times

---

[32] See Ithiel de Sola Pool, "Social Trends," *Science & Technology,* 76 (April 1968), pp. 87–101.

computers have been "misused," it may be that the seductiveness of a new tool is one aspect of the learning process associated with it. Experimentation and trial-and-error are essential aspects of research and learning. In some cases, a fascination with the possibilities of a new tool may lead to some experimentation which later appears to have been foolish. In other cases, experiments are undertaken with great expectations, only to be abandoned until some further advances in knowledge cause them to be taken up again with greater benefit. Language translation by computer might turn out to be an example of the latter. It is perhaps too soon to pass judgment on experimental uses of computers which may seem foolish in days to come. Some critics have argued that most uses of computers in humanistic research fall into this category. Yet computers have been used to great advantage in helping to build dictionaries, to decipher the characteristics of archaeological finds, and to analyze some historical documents.

In the natural sciences and, perhaps still less impressively, in the social sciences, computers have been used for both data analysis and problem solving. Simulation techniques and sophisticated manipulations of data have often generated unexpected or "serendipitous" results which have led to new discoveries. In this sense, computers have been what Anthony Oettinger has called creative "actors," rather than simply "instruments."

Computers have been actors in a more subtle sense, too. For the desire to exploit the possibilities of these tools has generated particular modes of research, *e.g.*, the gathering of large amounts of data, the use of various kinds of multivariate analysis. Some critics of contemporary social science contend that computers have thereby "corrupted" research in this field. Greater vehemence has been directed against the use of computers in research in the humanities. Insofar as some of the research that is being criticized is inadequately conceived or of poor quality, the researchers and not the computers are, of course, to blame. An undue and sometimes inappropriate desire to emulate the natural sciences is often responsible for the inappropriate use of computers.

The use of computers in the arts has often been a subject of either good-natured humor or ridicule. Computer created music, poetry, or drama has generally not been taken seriously. But in the process of this creation, some insight might be gained into the nature of art and what makes for great art.

Somewhat greater progress is being made towards an understanding of man's cognitive processes. Cybernetic research compares the functioning of the human brain with the functioning of computers. Although this branch of research is still in its early stages, it has generated much emotional reaction. Images have been evoked of "thinking machines" which are superior to man. Man's memory bank may be smaller than that of the computer, his ability to carry on several processes simultaneously is more limited, and the speed of his mental processes is not as great. Nevertheless, the cognitive processes that are built into computers are much simpler and better understood than those of the human brain.

Yet science fiction and popular imagery continue to portray the computer as a superbrain that is a threat to man. Perhaps it is necessary, as Bruce Mazlish has suggested, to sever the "discontinuity" between man and machine. Viewing machines as extensions of man might help to alleviate some of the distrust of technology. But this change in man's thinking and conception of himself might be as difficult to achieve as the breaking of the discontinuity between man and animal had been in the days of Darwin.

In some sectors of the population today, a deliberate attempt is made to do quite the reverse; man and machine are seen as radically discontinuous entities. If those who could not accept the continuity between man and animal often argued their case by dubbing man as "rational" and animal as "instinctual," there appears to be a tendency in some quarters today to see man as "emotional" and machine as "rational." Machines may execute tasks and may even "think," but they cannot "feel." So the argument runs. But it is an argument which is ultimately destructive, if one sees the abolition of this bifurcation between man's reason and his emotion as desirable.

CONCLUSION

It is impossible within a short space to do justice to the many ramifications of computer technology. Whole volumes could be written about the implications of computers in areas not even touched on in these pages, e.g., law enforcement, the practice of medicine, the stock market, international relations.

But it is not simply the scope of the subject which makes it so difficult. The rapidity with which both the technology and the society are changing militates against the possibility of making any definitive statements about the social implications of computers. Some of the issues are so current—e.g., the impact of computerized information systems on political decision-making—that it is not yet possible to discern the results. Other issues—e.g., the establishment of an electronic cash and credit system—are still futuristic.

All that one can hope for, therefore, is some greater understanding of the issues. It is clear that computers are powerful tools whose effects are quite profound. But like all tools, they are not autonomous. Their development and consequences are shaped by the social structures in which they operate.

The essays which follow are intended to present a broad sampling of the major social issues raised by computer technology. They have been selected to give the reader a sense of the concrete developments of computer technology and their implications in specific spheres of social activity. Highly speculative and futuristic essays have not been included; and in some areas of controversy, more than one essay on the topic has been reprinted. In an area of knowledge as new and difficult as this, however, it would be neither possible nor desirable to eliminate all controversy or speculation.

II

THE COMPUTER POTENTIAL

Introduction

To appreciate the impact of computers upon society requires some understanding of the kinds of operations that these machines perform. The selection by Thomas H. Crowley discusses the uses of the computer in providing economic benefits and in making possible certain types of data analysis. Some tasks that are now routinely performed would not have been feasible without the aid of the computer. The launching of satellites and the control of production lines, for example, are operations that require the results of computations to be available while the process is being carried on. In other cases, the computer may serve as an aid to judgment and decision-making by allowing the analyst to test the probable effects of certain courses of action. Thus an economist might program the computer with a model of the economy and test the effects of a tax cut. A businessman might use a computer to simulate what would happen if he were to raise his prices. Computer simulations also provide scientists and social scientists with insights into the nature of the processes that they study.

The utility of the computer is likely to be enhanced in the future. Harold Borko predicts that future computers will provide greater speed and storage capacity; smaller machines will be built, tailored to the needs of specific categories of users; character reading devices will be developed to convert data automatically into a computer code; and, eventually, programs will be constructed that will allow the computer to interpret and respond to natural language. These developments in computer technology will, in turn, help to stimulate the growth of other technologies by facilitating the work of the scientists and engineers upon whom such growth depends.

The brief selections by Herman Kahn and Anthony J. Wiener and by Ascher Opler serve to illustrate the diversity of opinion concerning the speed with which advances in computer technology are likely to occur. This type of dispute pervades the literature on the social effects of computers as well. Will computers revolutionize education, or medicine, or government operations in the next ten years or the next fifty? The answer often hinges not on the developments in computer technology *per se,* but on the nature of the institution in question—its organization, its finances, and its ability to overcome resistances to innovation and to make meaningful use of the new technology.

Methods and Uses

THOMAS H. CROWLEY

Bell Telephone Laboratories

There are many factors which determine whether a job is well suited to a computer, and to help simplify their consideration, we shall classify computer applications according to the primary reason for using a computer to do the job. Basically, a computer can be used for one, or more, of the following reasons: to gain economy, to make the job feasible at all, or to achieve insight into some process. We shall discuss each of these more fully by considering some examples.

As our first example, let us consider the preparation of the weekly payroll of a large company. The basic job is the preparation of a check for the appropriate amount of money. This amount of money is determined by the number of hours worked, the normal and overtime rate for this person, his payments for social security and income tax, and any other amounts to be withheld, e.g., hospitalization payments.

In order to prepare the check, weekly data is needed specifying the hours worked by each person on the payroll. In addition, the computer must have the following more-or-less fixed data available for each person: (1) name; (2) department or address; (3) hourly rate for normal time; (4) hourly rate for overtime; (5) number of income-tax exemptions; (6) social security paid so far this year; (7) amount of weekly hospitalization payment, if any. All this fixed information can be recorded on a master file which is usually a magnetic tape. It can be used over and over and thus does not have to be supplied every week on punched cards or some other relatively more expensive form of input. Once a week, therefore, the master file together with the weekly-hours-worked data (probably recorded on punched cards) serves as input data for the program which computes the amount of the check....

This program has at least two kinds of output. First, it may make some changes in the master tape; for example, if additional social security was withheld, it increases the amount withheld during this year. Second, it actually produces the checks. Generally this would be done by having some intermediate output (perhaps tape) generated which is used in turn to control a check-writing machine.

Although the information on the master file is relatively fixed, it does change occasionally; the man might be transferred to a different department, get a salary increase, or increase the number of his income-tax exemptions. Consequently, in addition to the actual wage computation another program is needed to edit the master file and bring all this information up to date. This change information is placed on punched cards and accumulated during each week. Shortly before the program to produce the weekly checks is run, this data is supplied to the edit program, and the master file is updated. It is then ready for use in the check computation.

The major objective of using the computer in such applications as these is simply economy. There are sometimes other objectives; e.g., fewer people are required; payroll production is faster and more accurate; but almost always the primary objective is to reduce the costs. . . .

One of the important characteristics of most applications of computers is discernible even in this brief description of an extremely over-simplified version of the job to be done. Although the task itself is simple—the program for computing the amount of the check requires only 10 or 20 instructions—there are many associated details which must be taken care of, e.g., reprogramming for tax-rate changes, obtaining information about changes of address, storing the tapes used for the master files, etc. Careful planning and execution of the entire process is essential if the system is to be successful since these associated "details" may well be controlling in determining how economically and efficiently the system functions.

This application illustrates some other important characteristics of one class of problems suited to computers. First, the amount of input data which must be manually prepared for each run is reasonable. If all the information on the master file had to be supplied manually each week, it is unlikely that there would be any savings. In other words, if the objective is economy, and it is more work to prepare the input data than it is to do the computation, don't do it on a computer! Second, much of the data, and the program itself, can be used over and over. Although this is a relatively simple program, the cost of writing it is much larger than the cost of preparing one of the weekly payrolls. It is only because the programming costs can be spread over many uses of the program that it is economical. Third, the problem itself involves many repetitions of a fairly simple job. This is almost a *sine qua non* of jobs suited to computers; if the job, or some substantial part of it, is not highly repetitious, then it probably is not economically done on a computer.

There are many industrial and business jobs with similar characteristics which are now being done on computers at greatly reduced costs. Preparing sales reports, controlling inventory, billing, preparing parts lists, and many others are all applications for which a major objective is economy. However, there are other applications which are characterized not by the resulting savings, but the fact that the jobs would not be *feasible* if they had to be done without using computers. Let us consider this class of applications next.

An excellent illustration of this type of application of computers is in the forecasting of weather. Because of its direct and obvious importance to man, predicting the weather (or formulating rules which attempt to minimize its adverse effects on such activities as crop planting) was probably one of the first technological questions to be attacked. Before the turn of the century, a mathematical model of the atmosphere including the equations which describe how the relevant variables such as air pressure, temperature, humidity, and velocity change over a period of time was formulated. At least as long ago as 1911, it was pointed out that the weather could be predicted by solving these equations. If the pressure, temperature, humidity, and velocity of the air are known for the same instant of time at many points around the world, the solution of the equations is mathematically quite straight-forward and accurately predicts what these quantities will be 1 hour, 1 day, or 1 week later. A network of weather stations around the world can make the measurements necessary to provide the initial conditions for these calculations, but it happens that a tremendous number of numerical calculations are required. Without a computer these computations required months of work, and there was obviously no point to predicting the weather, if the weather was already over! Consequently, this proposal lay essentially dormant until the advent of high-speed computers.

Immediately after it became clear that digital computers would vastly speed up such calculations, mathematicians and meteorologists intensified study of this method of prediction. Although the first results were crude, by 1962 a useful system was in routine operation. This system uses about 2,000 measurement reports to establish the initial conditions and then a large, high-speed computer requires 1 hour to make a 1-day prediction. More computation is required to make predictions over a longer period of time.

Although there is ample evidence that weather predictions are still not 100 percent accurate, these numerical predictions have increased the accuracy of certain forecasts quite significantly. However, the primary point of interest to us at this time is that this method of prediction was not even feasible before the advent of computers because the tremendous number of calculations required the very high speed of modern computers in order to obtain the results rapidly enough for them to be at all interesting.

A second important characteristic of this problem is that a very long sequence of calculations must be carried out with no mistakes. It would not do to make 100 million individual correct calculations (which is the order of magnitude involved in a 1-day forecast) and then to make a mistake on the last one and so predict rain on a sunny day, or vice versa. A computer is able to go through a very long error-free sequence whereas it is not possible for a person to take more than a few hundred or few thousand steps before making a mistake. In some applications it is only this high reliability of a computer which makes the job feasible at all.

A third, and distinct, characteristic may not be quite so obvious. In order to reduce the work involved in such calculations, a person would probably "round" the numbers to just a few significant figures and thus simplify the calculation. The computer is able to carry comfortably a very large number of figures so as to obtain very high precision. Another application may make this requirement more clear. When a missile is aimed at the moon, it is necessary to make extensive calculations to determine very small corrections to its course in order to fly close enough to take pictures. Each of these calculations must be carried out using a large number of significant figures in order to come close to a target almost 250,000 miles away.

To make the point again, computers are not used primarily as a means of reducing the cost of analytical weather forecasting. They are used because it is only with the high speed, reliability, and precision of the digital computer that this method is feasible at all.

To digress slightly, this is a good point at which to comment upon the frequently heard remark that "computers can do nothing that we cannot do ourselves." This statement is true only in a very trivial sense. Although, of course, it is possible to carry out exactly the same computations manually as are carried out by the computer, we don't do it because it is not worthwhile. R. W. Hamming has compared this statement with one that, regarded solely as a form of transportation, flying is no different than walking. Yet we all recognize that the drastic reduction in time required to get from one coast to the other via jet plane results in our making the trip many times, but we would never even consider walking the 3,000 miles. Similarly, the even greater increase in the speed and reliability of symbol processing by computer rather than by manual technique results in our doing things by computers that we would not do without them.

Many other jobs are feasible only with the aid of a computer because the output results must be available a very short time after the input data are available. These are called *real-time* applications since the computations must be made, and the results available, on the same time scale as that of the process involved. This contrasts with a job such as the payroll preparation in which the total computation time may well be a very small fraction of the week's time available for getting the results. The control of missile shots is a good example of a real-time computation. In order to put a satellite into a precise orbit, necessary instructions to the rockets, based on data obtained after liftoff, must be supplied at critical points during the launch. Only a computer can make the necessary calculations rapidly enough.

A somewhat more prosaic, but perhaps more important, example of a real-time application is the control of production lines. By continuously monitoring the items being produced, variations from the desired standard can be analyzed in a computer, and changes can be made in some of the production processes to reduce these variations. These are called *process-control* applications, and computers are now being used in this manner in oil refining, cement manufacturing, electronic parts manufacturing, and many other processes. Banking systems are also required to produce some information a very short time after the information is requested. For example, inquiries as to the balance in an account must be answered within minutes, and often a prodigious number of calculations are involved in keeping the balances correct.

Naturally, many computer applications are characterized both by being feasible only with a computer and also by leading to economies. For example, by using a computer to keep records of every item in the inventory of a factory, it may be feasible to reduce the size of the inventory to cover only one week's work without any appreciable danger of running out of a vital item and thus delaying production. Such a reduction in size of the inventory obviously results in savings.

A third major category of computer applications might be described by the phrase "computing for insight." Most of these applications are probably relevant primarily to scientists and engineers; however, one large group of examples, called *simulations*, are more generally used. For example, an economist who is trying to understand the interplay of a huge number of factors affecting our national economy may be able to create a mathematical model of the economy. In order to investigate the behavior of his model with certain real or assumed input data, he *simulates* the system on a computer. To do this he first produces a computer program which contains steps corresponding to each of the basic relations in his model. For example, his model might assume that a rise in the wages of steelworkers will be reflected in a corresponding increase in the wages of autoworkers. Then the program will contain a subroutine which assigns an increased wage to autoworkers if the input data indicate that steelworkers received an increase during this time period.

After the simulation program has been written, the economist tests it by supplying input data corresponding to real inputs and by allowing the program to run for a time long enough to simulate the operation of the economy during a prescribed period of real time. He can then analyze the behavior of the program to determine whether his model seems to be realistic, i.e., whether the calculated behavior agrees with what actually happened. If it is, he can then simulate the system with other assumed inputs to determine what the model says our economy might do under these conditions. For example, he might supply input data specifying a 10 percent cut in the corporate income tax with all other variables corresponding to actual figures for some year and try to determine what effect such a tax cut would have had.

Programs of this sort are being used to simulate the behavior of a business, or of an entire industry, or of a war! The construction of good models for these purposes is obviously very difficult; however, once a good model has been obtained, it may be possible to obtain more insight into the problem than could possibly be obtained in any other way. A businessman, for example, can simulate what would happen if he decided to double all his prices—if he actually tried it, he might go bankrupt. Simulations are also used to get information about new machines or new processes before they are operational. In addition to providing understanding, such simulations can have great economic value.

As one example of the many ways in which computers are used to gain scientific insight, we shall mention an application to an investigation of visual perception. An experimenter wanted to investigate some proposed models of how the human visual system determines spatial relations, and, in particular, he wanted to eliminate any part of this process which depended on previous knowledge. Such prior knowledge of familiar objects obviously could be important since, for example, when we see something that looks like a house, we have a pretty good idea of its spatial arrangement, no matter what we see. He decided to eliminate this possibility by using various "pseudo-random" patterns of dots as the objects to be examined by the subjects. In order to test various theories as to how binocular vision aids in the perception of depth, he used patterns made up of about 10,000 randomly determined dots; moreover, these patterns were constructed in pairs with one member of the pair having a portion of the pattern slightly displaced from the other. By using a computer to construct these carefully controlled pseudo-random patterns, he was able to test, and accept or eliminate, a number of conjectures about visual perception. These patterns would have been almost impossible to construct manually, and so such results would have been very difficult, if not impossible, to obtain without the use of the computer.

Looking back at [the] chronology of computer development. . ., we see that many of the needs which stimulated the early developments required many of the characteristics discussed in this chapter. For example, in the eighteenth century, large groups of clerks were employed to calculate, under the direction of a mathematician, very extensive tables of figures which were used for navigation. Following very explicit directions (a program!), the clerks carried out a highly repetitious series of additions and multiplications. Interestingly, it was found that the number of mistakes could be reduced if the clerks had very little mathematical training because they then followed their directions more readily and did not take shortcuts. However, even with very extensive checking it was impossible to eliminate mistakes— many of the mistakes were made in the final step of preparing the printed tables. This need for simple, accurate, and repetitive arithmetic operations and for reliable output stimulated Babbage to design his machine in an attempt to increase the reliability of these tables. In addition, of course, he expected the machine to be less expensive than the large groups of clerks.

Even earlier, weavers of patterned materials had come up against some of these same problems. Before 1750 they had originated the technique of punching holes in cards to specify a pattern. Thus, the first use of symbols stored in punched cards was apparently to represent instructions, not numbers! At first these instructions were simply interpreted by a man, or more likely a boy, and the process was slow, and mistakes were made. A series of modifications to the loom incorporated the punched card more mechanically into the process. The last step was made by Jacquard in 1790 when he produced a very successful loom in which all the power was supplied mechanically and all the control via punched card. Of course, the card for a particular pattern had to be prepared first by—shall we say—a programmer. In a restricted, but very important, sense the earliest programmable computers were used in process-control applications!

As another example, . . . Hollerith's work on punched-card machines was stimulated by his realization that it would not be feasible to analyze census data in the time available without some mechanical help.

During the last fifteen years, a continuous stream of new applications has been found. Computers have become faster, cheaper (in terms of equivalent computing capacity), and more reliable, and programmers have vastly improved automatic programming aids. We do not yet see any limits, except those of our own imaginations, to the uses which can be found for computers.

To summarize, we have classified the applications of computers into three categories according to their purpose—economy, feasibility, or insight. Naturally, these are frequently overlapping objectives, but at least one of them is necessary for any fruitful use of computers. One other point should be emphasized; although numbers played some role in most of the applications mentioned, the numbers are, in many ways, incidental. Computers are important today because they can manipulate arbitrary symbols and carry out complex procedures involving many decisions, thus producing economies, or theoretical insight, or doing jobs which would not be feasible without them.

Developments in Computer Technology

HAROLD BORKO

System Development Corporation

ADVANCES IN HARDWARE DESIGN

Greater Operating Speeds. Undoubtedly there will be more computers to-morrow than there are today, but what will be their salient characteristic? There are general trends in computer design, and if we [look at] the historical development and extend the trend line, we should be able to make some reasonable predications. The desire for greater operating speeds dominates both the early history and the more recent developments in computer technology. As progress was made, speed of computation was reduced from hours to minutes and then seconds. Now we deal with milliseconds and microseconds, and we already have a new word in our vocabulary—the nanosecond, or one-billionth of a second. Unquestionably, the computers of the future will be faster than the machines presently available.

Greater Storage Capacity. Concomitant with greater speeds, our extrapolated trend lines clearly indicate that the computers of the future will have greatly increased storage capacities. Progress is already being made in this direction...

Character Readers. Although progress has been made in increasing the internal operating speed and storage capacity of the computer, the methods of providing input data have remained relatively unchanged. Information is first coded and punched onto cards or paper tape. It may then be converted to a magnetic tape format or directly used as an input to the computer. In either case, converting information into a coded computer format is a slow and expensive process. The advent of character reader devices portends a

Reprinted from Harold Borko, ed. *Computer Applications in the Behavioral Sciences* (Englewood Cliffs, N.J.: Prentice-Hall, Inc., 1962), pp. 597–603, © System Development Corporation. Used with the permission of Prentice-Hall, Inc.

major breakthrough in this area. In essence, a character reading device is designed to convert data—words or numerals—into a computer code without human intervention. A number of such devices are currently available. Many banks are using magnetic character readers. Using this machine requires that the original information be printed in magnetic ink and in a special type font. Another type of character reading device is employed by the major oil companies for use with their credit cards. This device does not require the use of magnetic ink, but it does specify specially designed characters. Obviously, the next stage will be the design of equipment capable of scanning a page of ordinary type and automatically converting the information thereon to a code suitable for computer processing. Prototypes of such equipment have in fact been developed, and commercial models will be available within the next few years.

Greater Reliability of Operation. Currently, computers are used to process data which have been collected earlier. Thus two distinct steps are involved in the sequence of operations: first, the data are collected and coded; then, the data are processed. Since the two portions of the work are separated in time, the computer is always dealing with "old" data, such as yesterday's sales records. In most instances, this delay creates no hardship, because the results are still much more current than would have been possible without the use of the computer. There is, however, a growing trend to use the computer for "real-time" operational problems in which the data are processed as collected. The Air Defense SAGE System is an example of such use. In this application, the computer is used to monitor and update aircraft radar returns. But real-time operations are not limited to defense installations. The Remington-Rand UNIVAC 490 is a real-time system designed to provide business management with up-to-date operational information as it happens. This system is currently being used for airline reservation control, and it provides the airlines with accurate, instantaneous information of seat availability, cancellations, sales, and flight data.

The computer will be used to a greater extent than at present for real-time data processing. Before this prediction of future applications can come true, the computer must become a more reliable machine. In real-time operations, one cannot afford down time as a result of machine failure. This does not say that the present generation of computers is not reliable, but more time is spent on repairs and servicing than can be tolerated in economical real-time operations. Today's answer to this problem is found in duplexing the computer facilities, but this is an expensive solution. Future generations of computers will be more reliable and less subject to machine failure.

Some steps are already being taken in this direction, as can be inferred from the widespread use of transistorized printed circuit cards. These components are less subject to failure than are vacuum tubes. In the future—and the process has already started—the printed circuits will be manufactured in automated factories under computer control. Not only will automation cut costs, but component variability will be reduced and system reliability increased.

Change in Machine Logic. The very term "computer" implies the ability to calculate, i.e., to perform the standard arithmetic functions of addition, subtraction, multiplication, division, etc. Historically, this is what the machines were designed to do. But computers can do more, and the trend is to increase and improve upon their logical decision-making capabilities. At present, these improvements are being made within the framework of the traditional machine logic. But electronic data processing systems in the broadest sense of the term are not calculators but are symbol manipulators. In the future, machine logic will be designed to add numbers.

The first faltering steps in this direction are being taken now. Computers are being used to translate languages, to compose music, to simulate social organizations, etc. This trend will be accelerated as new logical components make their appearance.

Increased Development of Special Purpose Computers. In this area, the trend lines are not nearly as clear and defined as in the preceding areas. Nevertheless, present trends indicate that future designs will deemphasize the large expensive general purpose computers and will feature the smaller, more economical special purpose machines. These computers will be designed to do a particular job and to do it quickly, reliably, and economically.

In the early stages of development there is a great need for large, highly flexible machines. There is a need for experimentation in order to define the problem properly and to discover the means for solving it. There is a need for flexibility and the capability of doing many things—even though this capability is not exploited to its fullest. A high price is paid for these unused features. With maturity, the problems are defined and the operational procedures become standardized. Therefore, as the computer industry moves out of its early infancy to a more mature phase of development, there will be less need for general purpose machines. The data processing systems of the future will contain a number of compatible, special purpose computers, all integrated and operating with great efficiency.

ADVANCES IN PROGRAMING LANGUAGES

We have seen many advances in the design of computer hardware. Not the least of these is the proliferation of new computers that are appearing on the market at an increasing rate. This fact is of great significance to the programer. One can study, work hard, and become an expert programer on one machine. Then, for any one of a number of reasons (the programer takes a job with another company having a different machine, or his own company changes machines, etc.), he is no longer an expert. True, there is some transfer of training, but the programer must start again to learn new codes and a new machine language. Were only a single individual to find himself in this situation, it would be merely sad; but since this has been the common experience of many programers, the waste in manpower and money is tragic.

Problem-oriented Languages. A need exists for a compatible system in which programs can be readily translatable from one machine to another. It is highly unlikely, and somewhat illogical, to expect competing manufacturers to design compatible hardware. It is logical, however, to standardize the programing language, and a great deal of effort is being spent in this direction. Standardization can be accomplished only by changing the emphasis from programing in machine language to programing in a problem-oriented language. This change of emphasis is beginning to occur and the trend will be accelerated in the future. ALGOL (*algo*rithmic programing *l*anguage), designed for scientific numerical calculations, appeared in 1958. Concurrently, COBOL (*com*mon *b*usiness *o*riented *l*anguage) was developed as an aid in the solution of business data processing problems. In addition, many manufacturers and some large user organizations are developing their own compilers specifically applicable to their own equipment and functions.

These problem-oriented languages are helpful, but they cannot be regarded as the panacea for all programing difficulties. They are costly to build and require many man-years of labor to perfect. The effort is worthwhile if the investment can be amortized over a number of years of operational use. But, occasionally, the computer becomes obsolete soon after the compiler for it has been designed. What is needed is an intermediate language—intermediate between the problem-oriented language and the machine language. . . . Some of the nation's most skilled programers are working in this area, and it is predicted that their efforts will soon bear fruit.

Use of Natural Language. Looking now to a distant future, we can glimpse some even more exciting developments. Programing codes are very precise, formal language structures by which the operator communicates with the computer. Codes are necessary because the computer and the programer do not speak the same language. Until now we have directed our effort toward teaching men the language of the machine: i.e., they learned to write the instructions in machine language. The problem-oriented language is a compromise approach by which both men and machine learn a third, or intermediate, language. Why not go the whole way and teach the machine to understand the natural language of man?

Before this can be accomplished, progress must be made toward the solution of two major problems. The first has to do with pattern recognition—i.e., the development of hardware capable of sensing and recognizing the different patterns which make up the numbers and letters of the alphabet. . . . Progress is being made in this area. The second issue, and one more difficult to solve, involves programing the computer to interpret these sets of characters and to follow instructions or draw inferences. . . . Eventually. . .a computer program [will be developed] which will enable the machine to interpret and respond to instructions written in natural language.

Pursuing this line of reasoning further into the future, we raise the obvious question: "Why do we have to limit our communications to the computer to written messages?" If the computer can be taught to interpret written symbols, why can't it interpret spoken language? Bell Telephone

Laboratories and other research organizations are working in this area and are making good progress.

These are fascinating possibilities and they present some exciting challenges to research in general and programing in particular.

Developments in Computer Technology

HERMAN KAHN & ANTHONY J. WIENER

The Hudson Institute

Without elaborate input-output devices or sophisticated programs, a computer is really little more than a very large, very fast, and very complex abacus, but even so stripped, it has amazing potentialities. It has also had an amazing record of increase in potentiality over the last fifteen years. If one uses as a standard of measurement the size of the memory space divided by the basic "add time" of the computer (which measures roughly a computer's ability both to hold and to process information), then over the past fifteen years this basic criterion of computer performance has increased by a factor of ten every two or three years (this is a conservative estimate).

While some will argue that we will not duplicate this performance in the future because we are beginning to reach limits set by basic physical constraints, such as the speed of light, this may not be true, especially when one considers new techniques in time-sharing, segmentation of programs to add flexibility, and parallel-processing computers. . . .

The parallel-processing concept permits various elements of a complex problem to be solved simultaneously, rather than serially as in present systems. Other means of continuing the increase in computer capabilities by a factor of ten every several years may include using basic computational units that operate on the basis of matrices rather than single numbers; large-scale improvement in the present "soft-ware crisis" in programming language, treating complex operations as single units and combining them in both parallel and hierarchical operations, and so on. Thus even excluding the impact of new input-output devices and new concepts for programming and problem formulation, but just considering the basic capacity of the computer as a large and fast abacus has still meant that any doctrine about

Reprinted from Herman Kahn & Anthony J. Wiener. *The Year 2000: A Framework for Speculation* (New York: The Macmillan Company, 1967), pp. 88–89 by permission of the authors.

capabilities and limitations has had to be revised extensively every two or three years. But these exclusions are also important. About nine years ago, a program containing five thousand instructions was considered quite large. Now with the present capacities of computers and new programming languages...an individual can handle programs about ten times larger and a team may easily produce a program still larger by a factor of five to ten. These programs that used to take an hour or two to run now take a few seconds.

If computer capacities were to continue to increase by a factor of ten every two or three years until the end of the century (a factor between a hundred billion and ten quadrillion), then all current concepts about computer limitations will have to be reconsidered. Even if the trend continues for only the next decade or two, the improvements over current computers would be factors of thousands to millions. If we add the likely enormous improvements in input-output devices, programming and problem formulation, and better understanding of the basic phenomena being studied, manipulated, or simulated, these estimates of improvement may be wildly conservative. And even if the rate of change slows down by several factors, there would still be room in the next thirty-three years for an overall improvement of some five to ten orders of magnitude. Therefore, it is necessary to be skeptical of any sweeping but often meaningless or nonrigorous statements such as "a computer is limited by the designer—it cannot create anything he does not put in," or that "a computer cannot be truly creative or original." By the year 2000, computers are likely to match, simulate, or surpass some of man's most "human-like" intellectual abilities, including perhaps some of his aesthetic and creative capacities, in addition to having some new kinds of capabilities that human beings do not have. These computer capacities are not certain; however, it is an open question what inherent limitations computers have. If it turns out that they cannot duplicate or exceed certain characteristically human capabilities, that will be one of the most important discoveries of the twentieth century.

Developments in Computer Technology

ASCHER OPLER

Datamation Magazine

Few technical fields can match the computer world in its obsession with *time*. Our vocabulary is replete with terms like access time, nanosecond, latency, time slice, asynchronous activity, interrupt, time-sharing, simultaneity, overlap time, release date, clock pulse, scheduling algorithm, cycles, startstop time, purge date, etc.

Computer people are experts in dealing with time. Computer engineers analyze the status of each circuit on a clock pulse-by-pulse basis. Some programmers count timing cycles within tight inner loops, determine the number of milliseconds available to process an interrupt, etc. Schedulers plan computer operations so that waste time is minimized and high priority schedules are met.

However, when the computer specialist leaves the domain of computer time (picoseconds to hours) and enters the realm of human activity time (hours to years), he becomes a bumbling amateur. In the relatively brief history of electronic computers, an incredibly poor record for estimating time duration has been made. Although flagrantly bad estimates have drawn attention from time to time, not everyone realizes the breadth and depth of our errors in projecting the time required to achieve certain goals.

Serious failures have occurred in predicting both general and specific achievements. . . .

The history of the development of computer hardware, software and applications has been characterized by (1) lateness, (2) rescheduling. (3) cliffhanging finales, (4) substitution of interim versions for the promised ones, (5) the substitution of a "Phase 1" goal for the full goal, or (6) the on-time delivery of the promised system in a version whose quality and reliability were too poor to allow system usage.

Reprinted from Ascher Opler, "The Receding Future," *Datamation,* 13 (September 1967), pp. 31–32, by permission of *Datamation,* © F. D. Thompson Publications, Inc.

This sweeping condemnation has many well-known examples. It primarily applies more to the large, new and complex system than to the small, simple system that breaks no new ground. Acceptable operation of systems has frequently lagged six months behind schedule and occasionally three to four years.

Among the systems whose lateness embarrassed their sponsors were large business language compilers, command and control systems, real-time telecommunication systems, ultra-high speed computers, government-wide data processing systems, ultra-high capacity storage systems and large operating systems.

WHY ARE WE SO MYOPIC?

We understand what happens in one second of computer time; we do not understand what happens in one month of a man's time; we do not understand what happens in one year's time of technical development. If we are ever to leave the era of delayed fulfillment, we must improve our understanding of the true metric of time. Our vision of time is poor for a variety of reasons:

1. The time scale of the entire computer development is such that distortion arises easily. The technology of five years ago seems relatively primitive; that of 10 years ahead, "way out." In this environment, it is difficult to see two years ahead without distortion.

2. Marketing pressures in a highly competitive market are felt by everyone in the field. Under such pressure, one naturally makes claims for the earliest delivery of the most advanced system. Constant emphasis on *soonest, earliest, availability, delivery, installation* cannot help but prejudice all who estimate actual completion dates.

3. Lack of reliable industry experience on which to base our projecting. Chemical, automotive, steel and petroleum industries can make better projections based on many more years of experience. Experience in building computers, implementing software and real-time systems is diluted by the thousands of new people in the field and is rapidly made obsolescent by fast-changing technology.

4. So far we have failed to develop fully satisfactory methods for controlling the implementation of very large programs and, as a consequence, we have failed to develop accurate methods for predicting completion dates.

5. Underestimating the extent of development required to take a promising technical innovation and make it into a low-cost item usable with high reliability on an everyday basis. In other industries, a lag of at least five years is expected; in many fields, 10 years is normal. . . .

One hopeful trend is the gradual disillusionment with the products of crash programs and with accelerated early release versions. Bad experience after bad experience is hammering the lesson home. . . the reasonable user must wait a reasonable time to get a reasonable product.

III

THE ECONOMY

Introduction

The impacts of a new technology are generally felt first in the economic realm; the impact of the computer is no exception. The large-scale social ramifications of the computer result from its effects on the labor force, the process of production, and the way in which men transact business and receive services.

If it is hard to imagine how an economy of our size could have functioned without the automation of accounting and production processes, it is well to remember that over half of the labor force is employed in industries that are not automated.[1] Yet the computer has been an important force in bringing about the major economic changes which have occurred in recent decades: a shift from the predominance of the production of goods to the performance of services, the progressive elimination of the less skilled occupations, and the growth of professional and technical as well as white-collar work.

The concern so often expressed in the statement that "computers are throwing people out of work" does not appear to be justified. Clearly, the introduction of computers into factory or office does displace some workers; but it also generates new jobs and, by stimulating growth, it helps to maintain a high level of employment in the society. The National Commission on Technology, Automation, and Economic Progress has concluded that although technological change plays a major role in determining the particular workers who will be displaced, the rate of economic growth rather than technological change *per se* is the principal determinant of the general level of employment.[2]

Computers do, however, change the nature of the work that men do. In the fully automated, continuous-process industries, such as petrochemicals, the worker is freed from the drudgeries of the assembly line, but must exercise a higher degree of responsibility and control over his work. In the office, an increasing number of managers are using computers as aids in the decision-making process—although the extent of managerial reliance on

[1] See William A. Faunce, *Problems of an Industrial Society* (New York: McGraw-Hill, Inc., 1968), pp. 51–61.
[2] National Commission on Technology, Automation, and Economic Progress, *Technology and the American Economy* (Washington, D.C.: Government Printing Office, 1966).

the computer has probably been exaggerated[3]—and large numbers of office personnel are finding their work tied, directly or indirectly, to the computer.

While the amount of paper work has thus far grown faster than the introduction of electronic data processing, it is possible that in the future, positions in clerical and middle management work will decline. But jobs in information-processing and in the design and programming of the equipment will increase. The lower level jobs in the office are becoming more akin to those of the blue-collar workers in automated industries. "What have hitherto been manual jobs, albeit with a high degree of skill, have an increasing conceptual content, and an increasing emphasis upon formal knowledge. On other hand, some clerical jobs have an increasing manual content with the advent of computers."[4]

As the old blue-collar/white-collar distinction becomes less important, educational attainment and level of expertise increasingly become "the difference that divides."[5] Systems of "life-long learning" and some means of integrating job experience with educational experience will increasingly be necessary in the face of rapid knowledge obsolescence.

Computers are not only changing the nature of the work that men do, they are also altering the nature of some economic enterprises. The banking business, for example, has been changed since the wide-scale introduction of computers, so that such new functions as the handling of billing for physicians and dentists are becoming part of the bank's business. While computers have not yet had much impact in the services, aside from their use for routine administrative and accounting procedures, they are beginning to make some inroads here too. The entry of computers into the services provides a challenge to the professionals involved to learn how to use computers wisely. If computers are to be programmed to aid in medical diagnoses, physicians will have to be able to formalize the diagnostic process. If computerized instruction is to provide students with more than drill-and-practice exercise, educators will have to understand more about the learning process than they now do.

A more long-range implication of computers for the economy is the coordination and centralization that they often bring about. Within the corporation, "the ability to transmit, process, and analyze large masses of data, with virtually no delay and little regard for distances involved, permits the centralization of managerial controls in a single corporate location

---

3 In this connection, see "Unlocking the Computer's Profit Potential" (New York: McKinsey & Company, 1968); "Computer Management in Manufacturing Companies" (New York: Booz, Allen & Hamilton, Inc., 1967); and Rodney H. Brady, "Computers in Top-Level Decision Making," *Harvard Business Review,* 45 (July-August 1967), pp. 67–76.

4 Dorothy Wedderburn, "Are White-Collar and Blue-Collar Jobs Converging?" Document P 12–68 of the Third International Conference on Rationalization, Automation and Technological Change sponsored by the Metalworkers Industrial Union of the Federal Republic of Germany, (Oberhausen, Germany, 1968), p. 11.

5 John Kenneth Galbraith, *The New Industrial State* (Boston: Houghton Mifflin Co., 1967), p. 244.

rather than in a hierarchical range of regions and districts."[6] In the larger economic system, time-sharing systems for scientific research, for medical and legal work, and eventually, for a national electronic cash and credit system will generate more interdependence and hence a need for more regulation.

It should be noted, however, that the tendency towards centralization is not absolute or irreversible. Just as economic policy, rather than technological change *per se,* is the crucial factor in determining the level of employment in society, so do social factors, more generally, mediate between a technology and its effects. Thus the effects of the computer on the organization of the economy will depend on both the requirements of the technology and the structures and attitudes of the institutions into which it is introduced. While automation in the office, for example, often leads to centralization, it has been found that such centralization is not a necessary consequence of the technology.[7]

The selection by Paul Armer reviews the range and diversity of computer uses in industry and the services: in banking, credit and finances, publishing, agriculture, insurance, medicine, and education. In each case, both the present and potential impacts of the computer are described.

Herbert A. Simon examines the use of computers as an aid in managerial decision-making. He focuses on the possibility of programming non-routine decisions. Such automated decision-making, Simon argues, will not eliminate hierarchy or the division of responsibilities in the office. Although there is some movement towards decentralization in large United States business organizations, the design of data-processing and decision-making systems will tend to be a relatively centralized function, since the parts of the systems must mesh. The functions of middle management in expediting and setting the pace of work will probably recede in importance.

Albert A. Blum discusses the effects of computers on the skill level, job satisfaction, organization of work, and labor-union inclinations of white-collar workers. He observes that lower level routine jobs have been reduced in number, while higher level, more skilled clerical jobs have increased. But many of the highly skilled jobs resulting from automation will move out of the complex of clerical jobs and into managerial or professional levels. The middle level office jobs—those which require experience and seniority and some knowledge of company operations—are disappearing. Thus "automation appears to be cutting off the middle step in the old promotion ladder." Such reduction in the number of middle-level white-collar positions is, however, often caused less by automation directly, than by the re-organization of the office resulting from automation; and the organizational responses to automation may be variable.

Allen H. Anderson *et. al.* look at the problems and issues that will

---

[6] Boris Yavitz, "Technological Change," in Eli Ginzberg, ed., *Manpower Strategy for the Metropolis* (New York: Columbia University, 1968), p. 59.
[7] See Marshall W. Meyer, "Automation and Bureaucratic Structure," *American Journal of Sociology,* 74 (November 1968), pp. 256–264.

arise from the development of an electronic cash and credit system. The emergence of such a system, they argue, "appears imminent" and no major technological or economic breakthroughs are required. Questions immediately arise, however, as to who will own or control the system. The parties involved include banks, credit bureaus, retailers, hardware manufacturers, communications companies, finance companies, independent entrepreneurs, and the Federal Government. In addition, the institution of such a system would require a change in the financial habits of individuals—due to the elimination of time lags in processing checks, for example—and of the government—for example, in the amount and regulation of currency. An important prerequisite of the system is the existence of a sufficient supply of technical manpower.

The electronic cash and credit system would also depend upon the establishment of a computer utility to provide a communication system designed to operate between computers and between computers and people. Paul Baran examines the regulatory issues involved. Currently, he points out, the Federal Communications Commission must act as a protector of both the public and the utilities. It is therefore incapable of effectively regulating the emerging computer utility. Baran analyzes the current state of the industries involved and offers some policy suggestions. These include: initiating professional licensing standards for computer technicians so as to insure the privacy of users; allowing smaller computer utilities to work together so as to remove the economic advantage to large computer utilities; permitting the connection of computers to telephone systems; and encouraging the use of radio for data transmission.

Although recent declines in the length of the work year have not been spectacular, projections for the future indicate that leisure time will continue to grow. Erwin O. Smigel analyzes the different meanings of leisure and reviews research on the ways different groups in the population spend their leisure time. He then raises some questions as to the ability of Americans to cope with large amounts of free time.

FOR FURTHER READING

Robert Blauner, *Alienation and Freedom: The Factory Worker and His Industry* (Chicago: University of Chicago Press, 1964).

Rodney H. Brady, "Computers in Top-Level Decision Making," *Harvard Business Review,* 45 (July-August 1967), pp. 67–76.

Charles A. Myers, ed., *The Impact of Computers on Management* (Cambridge, Mass: M.I.T. Press, 1967).

National Commission on Technology, Automation, and Economic Progress, *Technology and the American Economy* (Washington, D.C.: Government Printing Office, 1966) and Appendix Volume II, *The Employment Impact of Technological Change.*

H.A. Rhee, *Office Automation in Social Perspective* (Oxford: Basil Blackwell, 1968).

Computer Applications in Industry and Services

PAUL ARMER

Stanford University

BANKING, CREDIT, AND FINANCIAL INFORMATION

The Present

Next to the Government, the commercial banking system is the largest processor of paper. In 1964, for example, the system handled 15 billion checks in addition to its other financial transactions. In the last 6 years, check handling has been largely automated, with over 90 percent of the checks in circulation today MICR-coded. (MICR stands for Magnetic Ink Character Recognition and is the scheme adopted by the banking industry for printing information on checks that can be read by character recognition machines.)

Almost all large banks have their own computers for check handling and many other applications. Many smaller banks have mechanized or are planning to do so through the utilization of computing services offered by a correspondent bank, computer service bureau, computer cooperative formed with other banks, or by the installation of sophisticated electronic bookkeeping machines. However, as an indication of the conservatism of some bankers, 45 percent of the banks (mostly small ones) replied to a 1962 questionnaire of the American Bankers Association that they had no intention of using a computer in the foreseeable future.[1]

Reprinted from Paul Armer, "Computer Aspects of Technological Change, Automation, and Economic Progress," in *The Outlook for Technological Change and Employment,* Appendix Volume I to *Technology and the American Economy.* Report of The National Commission on Technology, Automation, and Economic Progress (Washington, D.C.: Government Printing Office, 1966), pp. 218–220, 223–228 by permission of The RAND Corporation.

[1] *Automation and the Small Bank,* American Bankers Association, 1964.

Despite the introduction of computers, bank employment has continued to grow, thanks to the rapid growth of the industry as a whole. However, the rate of growth of employment has slowed down; and despite the overall growth, the bookkeeping function has been greatly affected by the introduction of computers. Some banks have reported reductions in their bookkeeping staffs of as much as 80 percent. For multibranch banks, the bookkeeping function has almost disappeared at the branch level. Despite this, however, the number of women employees as a percentage of total employment in banking dropped less than 1 percent from 1960 to 1964.[2] While bookkeepers have been disappearing, a new category of employee, associated with computer operation, has appeared.

Clearly, productivity has increased considerably in banking and will continue to do so. However, we lack a good measure of productivity and cannot say by how much it will increase.

In the short term, computer utilization will spread to most unautomated banks, and the number of applications in banks already using computers will increase. Total employment will continue to increase, although this will be of little solace to the displaced bookkeeper. New applications will result in increased productivity but will have little impact on employment.

Equipment trends will include increasing use of optical character-recognition equipment for documents other than checks and the introduction of teller terminals attached to a computer utility. For the larger banks, the computer utility will be inhouse, while the smaller banks will be served by an organization which serves a number of firms. Terminals will improve teller productivity, but the impact on employment will be masked by the growth of the industry.

Banks, particularly the large ones, will expand their services to include payroll, professional billing, account reconciliation, accounts receivable, inventory control, stock and bond portfolio analysis, bill collection, asset management, analyses of retail market penetration, economic forecasting, etc.—all essentially dependent on the use of a computer. Banks, thus, are already moving into the computer utility area, although for most existing applications the communications link remains the U.S. mail service or a courier. In California, however, the Bank of America offers a service to physicians and dentists whereby all charges and payments are reported each day over the telephone (augmented by a simple keyboard device) to the bank. The bank prepares monthly statements on its computer, mails them to the patients, and sends accounting reports to the physician or dentist. Typically, this system reduces accounting and billing time in the medical office by about 80 percent.

These new services not only affect the banking industry; they change the ways of handling business data and financial transactions in many other industries.

---

2 *Employment and Earnings Statistics for the United States, 1909–64*, December 1964, Bull. No. 1312–2, U.S. Department of Labor, Bureau of Labor Statistics, 1964.

The Future

> Automation affects not the mere mechanics of banking, but the very foundations of banking; not the individual bank, but banking systems and the national and international economies in which they are imbedded.
>
> —Anthony G. Oettinger[3]

The computer has forced the banking industry to examine itself to an unprecedented extent. Technological advances in computers and communications underscore the fact that banking is a system of national scope—in fact, worldwide scope.

The banking industry has observed that much of its activity could be eliminated, and there is a movement afoot to reduce drastically the paperwork in financial transactions—ultimately to do away with the check altogether. Other than cash, the simplest system would involve telling a financial computer utility via a store's terminal to transfer the amount of the sale from the buyer's to the store's account. If the purchaser's balance wouldn't cover the cost, the financial utility could extend him credit if his credit rating was good.

Many other less esoteric possibilities are already in use. For regular payments of a fixed amount, like mortage and insurance premiums or utility bills up to a given amount, the bill can be sent directly to the person's bank for payment. Other schemes involve using slightly augmented home telephones to instruct a bank to transfer funds to another account. (Such a system was demonstrated at the 1965 meeting of the American Bankers Association).

If businesses of all sizes have simple terminals linked to a central computer utility over a communications network, a universal credit-card system is possible. Various schemes could be used to make it difficult for someone else to use your "card"; e.g., "combination" key number known only to you; or ultimately, recognition by voice or thumbprint. Except for recognition by voice or print, all of the above is technologically feasible today. Pilot systems have been built and demonstrated, and economic feasibility is close at hand. But State and Federal laws will have to be changed and banks might have to agree to fundamental changes in systems and organization. For example, banks, savings and loan associations, and loan companies in a geographical area might establish a jointly owned and operated financial computer utility to serve the community.

Such a financial utility could develop a complete credit-deposit-loan history for each customer. This history could also enable the financial utility to be a more effective financial adviser to the customer, pointing out spending habits, making analyses, and helping with better financial analyses, and helping with better financial planning. Tax returns could also be turned out semiautomatically.

A financial utility could provide many other services; e.g., inventory control for sellers. Up-to-the-minute statistical information on other community activities could be made available.

---

[3] *Proceedings, National Automation Conference,* American Bankers Association. New York, July 13–16, 1964, p. 38.

In the system described above, money could be transferred from account to account or the transaction could involve the extension of credit. The credit could be given by the seller, based on an indication from the financial utility that the buyer was a good credit risk, or the credit could be extended by the utility. The latter scheme would probably mean less expensive credit for the buyer, since much of the cost of credit today goes toward the administrative costs of recordkeeping. Centralized in the financial computer utility, the costs of such recordkeeping would be lower.

No matter how credit is extended in the future, the credit information industry will undergo major changes in the next few years. Once again, existence of the computer is causing a close examination of present practices; e.g., the tremendous redundancy in information files on people and companies maintained in hundreds of places in a community, each with file clerks and credit managers. Not surprisingly, computerized credit bureaus are being established around the country: One is already in existence in Los Angeles and will soon cover all of California, and systems are also underway in New York and Texas. Needless to say, this portends the end of many small credit bureaus as well as the elimination of many jobs concerned with credit in companies which elect to buy this service. In California, for example, the large banks have signed up with the new credit information utility.

It is difficult to know how soon the financial information utility will come into being because the determining factors are social and political rather than economical or technological. Martin Greenberger suggests that it may be 25 to 30 years.[4] Rudolph Peterson, president of the Bank of America, predicts that we shall see a drop in the use of checks by the public in about 5 years. Dale L. Reistad, director of automation and marketing research for the American Banking Association, believes that citywide universal credit-card systems will soon come into existence and will be followed by areawide systems, finally forming a nationwide credit system by 1975.[5] The forces of the marketplace will operate to bring such systems into existence—where there are large savings, there are large markets and profits. (One company, Sperry Rand, has already developed the hardware for such a system and has demonstrated a pilot setup.)...[6]

THE PUBLISHING INDUSTRY

The impact of computers on publishing was very much in the news in 1965. The newspaper industry, in particular, became entangled in many labor-management problems directly as a result of automated composition proposals.

[4] Martin Greenberger, "Banking and the Information Utility," *Computers and Automation,* April 1965.
[5] Dale L. Reistad, "Banking Automation—1975." *Banking,* Journal of the American Bankers Association, July, October, and November 1964.
[6] *Consumer Purchasing Service,* Sperry Utah Co., Division of Sperry Rand Corp., Salt Lake City, Utah, 1965.

Automation in this industry dates back to the 1930's and involved typesetting equipment operated by teletype tape. Next the computer was added to produce the tape from material typed into the system, with the computer determining what should be on a given line, justifying and hyphenating as appropriate. Another step did away with hot metal through phototypesetting. The computer is also used in page makeup.

The latest developments are feasible both technologically and economically, and the rate of introduction will depend on labor agreements. The imminent obsolescence of the linotype operator is analogous with that of the railroad fireman.

Computers will be used in other areas of publishing: Material which is typed, edited, reworked, typed again, edited, reworked, and typed again will be typed into a computer from which it can be retrieved, edited, and reworked with a minimum of recopying. Simple systems of this kind, using a typewriter as a terminal, are in use today. Experimental systems using a television picture tube to display text also exist. Such systems not only reduce the amount of proofreading and typing but speed up the overall process. Once the material is in final form, the computer can feed it into a phototypesetter and produce a tape for later use or print the material for subsequent duplication. When image processing is added to the system, it will be possible to manipulate text and images simultaneously.

Information utilities may affect the publishing industry in another way: Some material that is printed today may be produced in smaller volume or possibly not printed at all. For example, books containing reference material that changes significantly in a short time might not be printed if the information utility can supply the information.

AGRICULTURE

Computer use in agriculture is in its infancy (less than 1 percent of U.S. farmers use such systems), but will probably become widespread in the next decade. Even small farmers, dealing with a computer service bureau by mail and eventually with a computer utility by telephone, can use computers profitably for recordkeeping, accounting and planning.

The computer will enable the farmer to:

Reduce clerical costs;
Know which operations are showing a profit;
Know which animals are paying their way;
Do a better job of financial management and reduce what he pays for credit;
Better manage labor, machinery, and other resources;
Determine "least-cost" rations for his animals;
Compare his performance with the average and best performance of other producers in his region.

Such EDP services are available from the agricultural extension services of a number of State universities and from several commercial organizations, including banks, with users reporting overall cost savings of from

5 to 20 percent with the more sophisticated systems. Thus computers will soon be contributing significantly to productivity in agriculture.

INSURANCE

The insurance industry, which deals primarily in information, naturally became an early user of electronic data-processing equipment. First applications were mechanizations of previous manual and punched-card processes. Integrated systems soon followed, often requiring a change in the organization of the company.

As a result of having entered the field early, the insurance industry has been almost completely penetrated by computers, and their major impact is now past. Most affected were the clerical workers. Yet, despite considerable increases in the productivity of the clerical force, the percentage of women employees in the total work force among insurance carriers fell only about 1 percent from 1958 to 1964.[7] Some continued increase in productivity in clerical operations can be expected. For example, the increased use of optical character recognition equipment will displace some keypunch operators and clerks.

Two developments in information processing will be important to the insurance industry in the future. The first is the introduction of the computer utility concept permitting terminals in the field throughout the country to be connected via a communications network to computer files in the home office. This will permit agents to give better and quicker service to their clients as well as reducing clerical costs. The second is the application of computers to underwriting, which is already being done by some companies, although only on policies of low face value and low risk. It is anticipated that this application will grow and reduce somewhat the need for underwriters.[8]

COMPUTERS AND HEALTH

The application of computers and information processing technology to health problems promises significant advances during the next decade in several ways:

1. The computer will take over much of the clerical work and information handling of those engaged in medical care;

2. Computer technology will provide more comprehensive information about each individual's medical history, resulting in better medical care;

3. Computer technology and automation will be applied to medical testing, resulting in cost reductions which, among other effects, will permit significant reductions in the costs of preventive medicine;

[7] *Employment and Earnings Statistics for the United States, 1909–64, op. cit.*
[8] Robert C. Goshay, *Information Technology in the Insurance Industry*, Richard D. Irwin, Inc., Homewood, Ill., 1964.

4. Computers will aid the physician in diagnosis;

5. Computers will be used to monitor the vital signs (e.g., blood pressure) of the seriously ill;

6. Computers as a research tool will aid in advancing medical knowledge;

7. Computers in the form of an information utility will aid in the dissemination of up-to-date knowledge to physicians.

Let us look at each of these applications in more detail:

1. As was often the case in industry, the first applications of computers in medical care involved the mechanization of manual procedures or punched card systems for such things as hospital payrolls and patient billing. Penetration of this phase of computer utilization into hospitals is far from complete, being limited to the larger and more progressive hospitals. Despite this limited penetration, there is already much interest in hospital information systems designed not only for accounting tasks but also to provide better medical care at a reduced price (with accounting information as a byproduct). Studies show that many hospital nurses spend as much as 40 percent of their time doing clerical work, and that test results and prescription orders may be transcribed as many as 10 times.

The hospital information system concept normally involves terminals at each nursing station and at such localities as admitting, pharmacy, medical records, laboratories, etc., connected to a central computer, which may be in the hospital or many miles away. Such a system would not only reduce the clerical load of administrative personnel, but also of doctors and nurses. Hospital care is one area where increased demand for services will rapidly absorb any increases in productivity. As a nation, we have been devoting more and more of our growing gross national product to medical care, while the number of professionally trained personnel per thousand population has been declining. Increased productivity is badly needed. Hospital employment has shown an average growth of 5.4 percent per year since 1958,[9] and will soon feel the added stimulus of the Medicare program.

2. One of the more important ways in which information systems will improve medical care will be through computer-based medical record systems containing the complete medical history of each individual in a geographical region. Such an information utility could be interrogated by a doctor or a hospital, even when the patient is far from home. Problems such as standards on terminology and data collection, deciding what is important to record and transmit, etc., will have to be solved before such systems can be made workable. But because of their obvious social value, we can anticipate that a large effort will be devoted to solving these problems and developing less ambitious systems. Large amounts of medical data in computer-processable form will also be of great value to medical researchers.

3. Much effort is presently being devoted to automating various tests

---

9 *Employment and Earnings Statistics for the United States, 1909–64 op. cit.*

in the interest of improving medical care and reducing costs. Hopefully, the costs of a number of important tests can be reduced to the point where they can be routinely given to everyone in a program of preventive medicine. For example, even if we could presently afford to administer electrocardiograms to everyone annually, there are not enough doctors to analyze them. Ample evidence indicates that computers could at least select those electrocardiograms requiring further examination by a cardiologist and could probably do an adequate job of classifying abnormal cases.[10] Cost reductions of a factor of 10 appear quite feasible for many tests.

Two North Carolina hospitals recently automated a laboratory where 11 chemical tests are routinely run on the blood of patients, as contrasted with the usual 2 standard determinations. Unexpected data of direct benefit to the patient were found in 1 out of every 15 admissions.[11]

Another example of the use of computers in a program of preventive medicine exists at the Kaiser Foundation Health Plan in the San Francisco Bay area. On entering the clinic the patient is given a self-administered questionnaire of some 600 questions about his medical status. His answers and the results of a number of tests (blood tests, electrocardiogram, etc.) are fed into a computer which makes a "provisional diagnosis" that is used by physicians in subsequent examinations.

4. Although computer-aided diagnosis is primarily confined to research efforts at present, the results give promise of providing a useful operational tool for the physician in the next decade.[12] The computer will not replace the judgment of the physicians, but rather will be a valuable aid to him in arriving at a diagnosis, much like a consultant who suggests one or more tentative diagnoses. The computer process could also suggest tests which would enable the physician to decide among several tentative diagnoses or to confirm the one tentatively arrived at. In most instances the physician would undoubtedly have called for the same tests, but in these days of advanced medical science, there are many diseases and medical conditions which a doctor seldom encounters.

More important, in a program of preventive medicine, the computer in the 1970's should be able to digest facts about the present medical status of individual patients (the majority of whom are healthy) and separate out those cases warranting the further attention of a doctor. The computer process would intentionally be conservative; if there were any doubt, consultation with a physician would be indicated. While it is true that most patients might not be seen by a doctor under such a system, at the present

10 H. R. Warner, A. F. Toronto, and L. G. Veasy. "A Mathematical Approach to Medical Diagnosis. Application to Congenital Heart Disease." *J.A.M.A.*, vol. 177. 1961, pp. 177–183: and H. R. Warner, A. F. Toronto, and L. G. Veasy, "Experience with Bayes' Theorem for Computer Diagnosis of Congenital Heart Disease." *Ann. N.Y. Acad. Sci.*, vol. 115. 1964, pp. 558–567.

11 John A. Osmundsen, "Automation Used for Blood Tests," *New York Times*, Sept. 12, 1965.

12 John A. Jacquez (ed.), *Proceedings of a Conference held at the University of Michigan, May 9–11, 1963*, Ann Arbor, Mich., 1964.

time most apparently healthy people in the United States do not receive periodic medical checkups at all. Thus, we have here another situation in which society can readily absorb the increased productivity brought about by computers, automation, and technological progress.

5. Computers capable of monitoring such physiological variables as pulse, blood pressure, and temperature have been much discussed in recent years, and experiments have been carried out in intensive care units.[13] One expert has estimated that as many as 10 heart deaths out of every 100 could be prevented if patients were in an intensive coronary care unit where electro-cardiograms and pulse rate were continuously monitored.[14]

The use of computers to monitor these variables will undoubtedly gain wide acceptance in intensive care units, but will probably not be applied generally in hospitals during the next decade because of high costs and the fact that the measuring instruments tend to be uncomfortable and bothersome.

6. Computers are used extensively today in medical research centers, and their use as research tools will undoubtedly grow rapidly. They are used to collect and analyze experimental data, sometimes with the computer controlling the experiment based on immediate analysis of the information being collected. An increasingly important use of computers as a research tool involves the construction of mathematical models that embody the essence of a researcher's theory about the functioning of a biological system. Such models not only permit the researcher to study complex biological systems but are often useful in teaching medical students.[15]

7. A recent World Health Organization review of computer use in medicine reported that there is a great need for an "electronic encyclopedia" of medical knowledge.[16] At present, the physician faced with a case difficult to diagnose generally seeks the advice of colleagues and consults textbooks; but his colleagues may not have any more extensive knowledge than he does, and the textbooks may be out of date. A medical information utility, however, could make the most recent information available.

COMPUTERS AND EDUCATION

When Sputnik jarred national pride in 1957, the United States stepped up its commitment to education, and more recently has embarked on even

---

[13] Robert I., Patrick and Marshall A. Rockwell, Jr., "Patients On-Line," *Datamation,* vol. 11. No. 9, September 1965, pp. 57–60.

[14] John A. Osmundsen, "Electronic Units Aid Heart Patient," *New York Times,* Oct. 16, 1965.

[15] J. V. Maloney, Jr., M.D., J. C. DeHaven, E. C. DeLand, and G. B. Brandham. M.D., *Analysis of Chemical Constituents of Blood by Digital Computer,* The RAND Corp., April 1963.

[16] M. S. Wilde, "Computers to Furnish World-Wide Diagnoses," *Los Angeles Times,* Oct. 7, 1965.

greater efforts. Education now engages over 5 percent of the labor force (excluding students),[17] and costs in excess of $30 billion a year.[18]

As school-age population increases, as the average educational level rises, and as the philosophy of viewing education as a continuous process throughout our lifetimes is adopted, the Nation's commitment to education must continue to grow.

Most educators and psychologists agree that the present educational process is deplorably less efficient than it might be. Clearly, the computer is destined to play an important role in improving efficiency during the next decade in several ways:

1. The computer will take over much of the clerical work and information handling in education;

2. Computer technology, with other technological developments, will increase the student's productivity by permitting individualized instruction. This is potentially much more important than (1) above;

3. Computers wll become important tools for educational research and development and will help psychologists and educators to a better understanding of the learning process.

Let us look at each of these in more detail:

1. Since education has all the same management and administrative problems as industry, plus a few more, computer and information technology is applicable; and while education lags behind most of industry in computer utilization, it has begun to adopt computers in a major way. This will result in increased productivity among administrators as well as teachers and students and in better utilization of capital resources. As mentioned earlier, using a computer for classroom scheduling creates classroom space.[19] As in industry, increased productivity from managers and clerical workers in education is to be expected; but the impact on teachers and students is possibly not so obvious. Teachers, too, spend much time in clerical activities. For example, the Richmond (California) pilot study found that computerizing the reporting of grades added at least 4 effective teaching days to the school year.[20] Technological and economic feasibility exists for these applications today, and it is only a matter of time until they are adopted.

We can also contemplate the possibility of a terminal connected to a central computer for each teacher to aid with information processing. Such systems will soon be technologically feasible, but economics will probably delay widespread use of teacher terminals until at least the early 1970's.

2. Many of the inefficiencies of present educational systems are connected with the fact that instructors must usually deal with their class as groups and not with individual students. Should the level of teaching be geared to the slowest or the fastest learners? The great hope is that com-

[17] *Employment and Earning Statistics for the United States, 1909–64, op. cit.*
[18] Gilbert Burck, "Knowledge: The Biggest Growth Industry of Them All," *Fortune,* November 1964, pp. 128–131, 267–270.
[19] "School Scheduling," *op. cit.*
[20] *A Report of An Experiment—The State Pilot Project in Educational Data Processing.* Monograph No. 3, Educational Systems Corp., Malibu, Calif., 1964.

puters and technology will permit individualized instruction so that each student can work at his own pace and not proceed to new material until he has mastered the old. Not only will individualized instruction help the slow learners, but equally important, we may realize the real potential of the gifted child. In spite of all the efforts to recognize and help the gifted, much more time is devoted to below-average children.

Individual instruction is usually based on notions of programed learning. This is a controversial subject: programed learning has not, to date, lived up to its advance publicity, and simple teaching machines have been an economic failure.[21] But students can learn from programed materials; and as techniques improve, particularly with the added scope and greater sophistication permitted by the use of a computer (the system envisaged consists of student terminals with a display device and a keyboard), researchers have great expectation for programed instruction.

While some areas of instruction can be readily adapted to the computer, others are much more difficult. For example, drill-and-practice systems are comparatively easy to program. . . .

Tutorial systems, where the aim is to develop the pupil's skill in using a given concept, are somewhat more difficult than drill-and-practice systems. . . .

Dialogue systems, in which discourse between the student and the computer system would take place in the same fashion as exists between student and teacher, will obviously be much harder to develop. . . .

Although some areas of computer-aided instruction (CAI) are technologically feasible today, economic feasibility is another matter. For some time, CAI will be confined to experimental research and to specific areas of adult education where the cost can be justified (e.g., electronic technician training). Widespread use will depend not only on research progress, but also on the size of the financial commitment the Nation makes to education.

Individualizing instruction depends on knowing what material the individual has mastered. CAI systems must determine a student's status based on an analysis of his responses to the computer. This assessment of the student's progress determines the material to be presented next. Without CAI, the teacher will lack the time to develop and give remedial instruction to each student. However, the computer could, based on the analysis of the difficulties, instruct the student to review certain portions of the text.

3. The computer is a useful tool in the performance of research and development in education as it is in almost all R. & D. work. It permits the researcher to deal with more complex data and greater volumes of information than he could otherwise. CAI systems will be able to capture and preserve much greater quantities of data than were previously available.

Computers are being increasingly used in psychological research, both to analyze experimental data and to develop psychological theories about information processes in humans. The behavior of a computer is determined

[21] Kenneth O. May, *Programed Learning and Mathematical Education*, a CEM study, Committee on Educational Media, Mathematical Association of America, 1965.

by a program, and the goal is to develop a computer program which causes it to perform an information-processing task the same way that people do. The program is a model which represents the researcher's hypotheses about the information processes underlying human cognition, just as the computer program which calculates missile trajectories represents a theory of the flight of a missile. The implications of the model are determined by running the program on the computer with varying inputs and observing the outputs.[22]

The implications for improved efficiency in our educational system stemming from a better understanding of the human learning process are tremendous, whether that understanding comes from computer modeling of human cognition or from some other research strategy.

Information Processing Education

Instruction in the science of information processing itself is another important area. Since computers are playing an increasingly important role in society, it is vital that a large segment of our population understand information processing and know how to utilize computer power. Not that everyone must be a computer programer any more than we have had to become automobile mechanics, but understanding and knowing how to use the computer are important. Numerous experiments have established that computing can be learned at an early age, say the 11th or 12th grade level. Indeed, it can be argued that it is best learned at that level. However, our secondary school instructors are unprepared to teach the subject.

The computer is already a powerful tool for the college student, and the trend is clearly toward having college students learn its use. Some 500 college campuses have at least one computer. In many schools—particularly engineering schools—the students use computers for appropriate assignments. They do their own programing, punch their own input cards, and leave the punched cards at the computing center on their way to class, retrieving the output an hour or so later or the next day. Soon, computer terminals will exist at various places around the campus at many universities; several schools already have this capability. It is implicit in such a setup that the course assignments require a computer to complete them: hand calculation methods would swamp the students. The problems assigned can be realistic, rather than oversimplified models in which the answers "come out even." Oddly enough, there is not a single text—at any level or in any subject— that assumes student use of this new capability, but they are surely to come.

The movement of computers into the schools has been slow and haphazard, but it will be speeded up as low-cost but powerful machines and time-shared terminals offer the schools economical and vast computing power.

---

22 E. A. Feigenbaum, and Julian Feldman (eds.), *Computers and Thought,* McGraw-Hill, New York, 1963, pp. 269–386.

Management and Decision-Making

HERBERT A. SIMON

Carnegie-Mellon University

In discussing how executives now make decisions, and how they will make them in the future, let us distinguish two polar types of decisions. I shall call them *programmed decisions* and *nonprogrammed decisions,* respectively. Having christened them, I hasten to add that they are not really distinct types, but a whole continuum, with highly programmed decisions at one end of that continuum and highly unprogrammed decisions at the other end. We can find decisions of all shades of gray along the continuum, and I use the terms programmed and nonprogrammed simply as labels for the black and the white of the range.[1]

Decisions are programmed to the extent that they are repetitive and routine, to the extent that a definite procedure has been worked out for handling them so that they don't have to be treated *de novo* each time they occur. The obvious reason why programmed decisions tend to be repetitive, and vice versa, is that if a particular problem recurs often enough, a routine procedure will usually be worked out for solving it. Numerous examples of programmed decisions in organizations will occur to you: pricing ordinary customers' orders; determining salary payments to employees who have been ill; reordering office supplies.

Decisions are nonprogrammed to the extent that they are novel, unstructured, and consequential. There is no cut-and-dried method for handling the problem because it hasn't arisen before, or because its precise nature and structure are elusive or complex, or because it is so important that it deserves a custom-tailored treatment. . . .

Reprinted from Herbert A. Simon. *The Shape of Automation for Men and Management* (New York: Harper & Row, Inc., 1965), pp. 58–59, 75–76, 81–83, 85, 90–92, 98–109, Copyright © by School of Commerce, Accounts, and Finance, New York University. Used by permission of Harper & Row, Publishers.

[1] See James G. March and Herbert A. Simon, *Organizations* (New York: John Wiley & Sons, 1958), pp. 139–142 and 177–180 for further discussion of these types of decisions. The labels used there are slightly different.

The revolution in programmed decision making has by no means reached its limits, but we can now see its shape. The rapidity of change stems partly from the fact that there has been not a single innovation but several related innovations, all of which contribute to it.

1. The electronic computer is bringing about, with unexpected speed, a high level of automation in the routine, programmed decision making and data processing that were formerly the province of clerks.

2. The area of programmed decision making is being rapidly extended as we find ways to apply the tools of operations research to types of decisions that have up to now been regarded as judgmental—particularly, but not exclusively, middle-management decisions in the area of manufacturing and warehousing.

3. The computer has extended the capability of the mathematical techniques to problems far too large to be handled by less automatic computing devices, and has further extended the range of programmable decisions by contributing the new technique of simulation.

4. Companies are just beginning to discover ways of bringing together the first two of these developments: of combining the mathematical techniques for making decisions about aggregative middle-management variables with the data-processing techniques for implementing these decisions in detail at clerical levels.

Out of the combination of these four developments there is emerging the new picture of the data-processing factory for manufacturing, in a highly mechanized way, the organization's programmed decisions—just as the physical processing factory manufactures its products in a manner that becomes increasingly mechanized. The automated factory of the future will operate on the basis of programmed decisions produced in the automated office beside it.

HEURISTIC PROBLEM SOLVING[2]

However significant the techniques for programmed decision making that have emerged over the last decade, and however great the progress in reducing to sophisticated programs some areas that had previously been unprogrammed, these developments still leave untouched a major part of managerial decision-making activity. Many, perhaps most, of the problems that have to be handled at middle and high levels in management have not been made amenable to mathematical treatment, and probably never will.

[2] This section is based mainly on research on complex information processing sponsored by the Graduate School of Industrial Administration at Carnegie Institute of Technology and by The RAND Corporation, in which I have been engaged with my colleagues, Allen Newell and J. C. Shaw. Most of the ideas in it are our joint product. See Newell, Shaw, and Simon, "The Elements of a Theory of Human Problem Solving," *Psychological Review*, vol. 65, March 1958, pp. 151–166; Newell, Shaw, and Simon, "A General Problem Solving Program for a Computer," *Computers and Automation*, vol. 8, July 1959, pp. 10–17; and Newell and Simon, "Heuristic Problem Solving," *Operations Research*, vol. 6, January-February 1958, pp. 1–10, and *ibid.*, May-June 1958, pp. 449–450.

But that is not the whole story. There is now good reason to believe that the processes of nonprogrammed decision making will soon undergo as fundamental a revolution as the one which is currently transforming programmed decision making in business organizations. Basic discoveries have been made about the nature of human problem solving. While these discoveries are still at the stage of fundamental research, the first potentialities for business application are beginning to emerge. We may expect this second revolution to follow the first one, with a lag of ten to twenty years. . . .

In solving problems, human thinking is governed by programs that organize myriads of simple information processes—or symbol manipulating processes if you like—into orderly, complex sequences that are responsive to and adaptive to the task environment and the clues that are extracted from that environment as the sequences unfold. Since programs of the same kind can be written for computers, these programs can be used to describe and simulate human thinking. In doing so, we are not asserting that there is any resemblance between the neurology of the human and the hardware of the computer. They are grossly different. However, at the level of detail represented by elementary information processes, programs can be written to describe human symbol manipulation, and these programs can be used to induce a computer to simulate the human process. . . .

The processes of problem solving are the familiar processes of noticing, searching, modifying the search direction on the basis of clues. The same elementary symbol-manipulating processes that participate in these functions are also sufficient for such problem-solving techniques as abstracting and using imagery. The secret of problem solving is that there is no secret. It is accomplished through complex structures of familiar simple elements. The proof is that we can simulate it, using no more than those simple elements as the building blocks of our programs.

From the standpoint of human simulation, perhaps the most interesting program of this kind is one labeled GPS (General Problem Solver).[3] It is called GPS not because it can solve any kind of problem—it cannot—but because the program itself makes no specific reference to the subject matter of the problem. . . .

GPS is a program—initially inferred from the protocols of human subjects solving problems in the laboratory, and subsequently coded for computer simulation—for reasoning in terms of ends and means, in terms of goals and subgoals, about problematic situations. It is subject matter free in the sense that it is applicable to any problem that can be cast into an appropriate general form (e.g., as a problem of transforming one object into another by the application of operators). It appears to reproduce most of the processes that are observable in the behavior of the laboratory subjects and to explain the organization of those processes. On the basis of simulation, we can say that GPS is a substantially correct theory of the problem-solving process as it occurs under these particular laboratory conditions. How general it is remains to be seen. . . .

[3] For a fuller description of the General Problem Solver, which was developed by the Carnegie-RAND research group, see the second reference cited in footnote 2.

Success in simulating human problem solving can have two kinds of consequences: It may lead to the automation of some organizational problem-solving tasks; it may also provide us with means for improving substantially the effectiveness of humans in performing such tasks. Let us consider these two possibilities and their inter-relations.

If I am right in my optimistic prediction that we are rapidly dissolving the mysteries that surround nonprogrammed decision making, then the question of how far that decision making shall be automated ceases to be a technological question and becomes an economic question. Technologically, it is today feasible to get all our energy directly from the sun, and to be entirely independent of oil, coal, or nuclear fuels. Economically, of course, it is not feasible at all. The capital investment required for direct conversion of the sun's rays to heat is so large that only in a few desert climates is the process even marginally efficient.

Similarly, the fact that a computer can do something a man can do does not mean that we will employ the computer instead of the man. Computers are today demonstrably more economical than men for most large-scale arithmetic computations. In most business data-processing tasks they are somewhere near the breakeven point—whether they can prove themselves in terms of costs depends on the volume of work and on the biases of the man who makes the calculations. As chess players, they are exceedingly expensive (quite apart from the low quality of their play at the present time.)

To put the matter crudely, if a computer rents for $10,000 a month, we can not afford to use it for nonprogrammed decision making unless its output of such decisions is equivalent to that of ten men at middle-management levels. Our experience to date—which is admittedly slight—suggests that computers do not have anything like the comparative advantage in efficiency over humans in the area of heuristic problem solving that they have in arithmetic and scientific computing.

There is little point in a further listing of pros and cons. As computer design evolves and as the science of programming continues to develop, the economics of heuristic problem solving by computer will change rapidly. As it changes, we shall have to reassess continually our estimates as to which tasks are better automated and which tasks are better put in the hands—and heads—of the human members of organizations. About the only conclusion we can state with certainty is that the boundary between man and computer in data processing work has moved considerably in the past five years, and will almost surely continue to move. . . .

ORGANIZATIONAL DESIGN: MAN-MACHINE SYSTEMS FOR DECISION MAKING

With operations research and electronic data processing we have acquired the technical capacity to automate programmed decision making and to bring into the programmed area some important classes of decisions that were formerly unprogrammed. Important innovations in decision-making processes in business are already resulting from these discoveries.

With heuristic programming, we are acquiring the technical capacity to automate nonprogrammed decision making. The next two decades will see changes in business decision making and business organization that will stem from this second phase in the revolution of our information technology. I should like now to explore, briefly, what the world of business will look like as these changes occur. . .[4]

An organization can be pictured as a three-layered cake. In the bottom layer, we have the basic work processes—in the case of a manufacturing organization, the processes that procure raw materials, manufacture the physical product, warehouse it, and ship it. In the middle layer, we have the programmed decision-making processes, the processes that govern the day-to-day operation of the manufacturing and distribution system. In the top layer, we have the nonprogrammed decision-making processes, the processes that are required to design and redesign the entire system, to provide it with its basic goals and objectives, and to monitor its performance.

Automation of data processing and decision making will not change this fundamental three-part structure. It may, by bringing about a more explicit formal description of the entire system, make the relations among the parts clear and more explicit.

THE HIERARCHICAL STRUCTURE OF ORGANIZATION

Large organizations are almost universally hierarchical in structure. . . . Hierarchical subdivision is common to virtually all complex systems of which we have knowledge. . . .

The near universality of hierarchy in the composition of complex systems suggests that there is something fundamental in this structural principle that goes beyond the peculiarities of human organization. I can suggest at least two reasons why complex systems should generally be hierarchical:

1. *Among possible systems of a given size and complexity, hierarchical systems, composed of subsystems, are the most likely to appear through evolutionary processes.* A metaphor will show why this is so. Suppose we have two watchmakers, each of whom is assembling watches of ten thousand parts. The watchmakers are interrupted, from time to time, by the telephone, and have to put down their work. Now watchmaker A finds that whenever he lays down a partially completed watch, it falls apart again, and when he returns to it, he has to start reassembling it from the beginning. Watchmaker B, however, has designed his watches in such a way

---

[4] See H. J. Leavitt and T. L. Whisler, "Management in the 1980's," *Harvard Business Review,* vol. 36, no. 6 (November-December 1958), pp. 41–48; and H. A. Simon, "The Corporation: Will It Be Managed by Machines?" paper prepared for *Management and Corporations, 1985,* tenth anniversary symposium, Graduate School of Industrial Administration, Carnegie Institute of Technology, April 21, 1960 (McGraw-Hill).

that each watch is composed of ten subassemblies of one thousand parts each, the subassemblies being themselves stable components. The major subassemblies are composed, in turn, of ten stable subassemblies of one hundred parts each, and so on. Clearly, if interruptions are at all frequent, watchmaker B will assemble a great many watches before watchmaker A is able to complete a single one.

2. *Among systems of a given size and complexity, hierarchical systems require much less information transmission among their parts than do other types of systems.* As was pointed out many years ago, as the number of members of an organization grows, the number of *pairs* of members grows with the square (and the number of possible subsets of members even more rapidly). If each member, in order to act effectively, has to know in detail what each other member is doing, the total amount of information that has to be transmitted in the organization will grow at least proportionately with the square of its size. If the organization is subdivided into units, it may be possible to arrange matters so that an individual needs detailed information only about the behavior of individuals in his own unit, and aggregative summary information about average behavior in other units. If this is so, and if the organization continues to subdivide into suborganizations by cell division as it grows in size, keeping the size of the lowest level subdivisions constant, the total amount of information that has to be transmitted will grow only slightly more than proportionately with size.

The reasons for hierarchy go far beyond the need for unity of command or other considerations relating to authority.

The conclusion I draw from this analysis is that the automation of decision making, irrespective of how far it goes and in what directions it proceeds, is unlikely to obliterate the basically hierarchical structure of organizations. The decision-making process will still call for departmentalization and subdepartmentalization of responsibilities. There is some support for this prediction in the last decade's experience with computer programming. Whenever highly complex programs have been written—whether for scientific computing, business data processing, or heuristic problem solving—they have always turned out to have a clear-cut hierarchical structure. The over-all program is always subdivided into subprograms. In programs of any great complexity, the subprograms are further subdivided, and so on. Moreover, in some general sense, the higher level programs control or govern the behavior of the lower level programs, so that we find among these programs relations of authority among routines that are not dissimilar to those we are familiar with in human organizations.[5]

Since organizations are systems of behavior designed to enable humans and their machines to accomplish goals, organizational form must be a joint function of human characteristics and the nature of the task environment. It must reflect the capabilities and limitations of the people and tools

[5] The exercise of authority by computer programs over others is not usually accompanied by effect. Routines do not resent or resist accepting orders from other routines.

that are to carry out the tasks. It must reflect the resistance and ductility of the materials to which the people and tools apply themselves. What I have been asserting, then, in the preceding paragraphs is that one of the near universal aspects of organizational form, hierarchy, reflects no very specific properties of man, but a very general one. An organization will tend to assume hierarchical form whenever the task environment is complex relative to the problem-solving and communicating powers of the organization members and their tools. Hierarchy is the adaptive form for finite intelligence to assume in the face of complexity.

The organizations of the future, then, will be hierarchies, no matter what the exact division of labor between men and computers. This is not to say that there will be no important differences between present and future organizations. Two points, in particular, will have to be reexamined at each stage of automation:

1. What are the optimal sizes of the building blocks in the hierarchy? Will they become larger or smaller? This is the question of centralization and decentralization.

2. What will be the relations among the building blocks? In particular, how far will traditional authority and accountability relations persist, and how far will they be modified? What will be the effect of automation upon subgoal formation and subgoal identification.

Size of the Building Blocks:
Centralization and Decentralization

One of the major contemporary issues in organization design is the question of how centralized or decentralized the decision-making process will be—how much of the decision making should be done by the executives of the larger units, and how much should be delegated to lower levels. But centralizing and decentralizing are not genuine alternatives for organizing. The question is not whether we shall decentralize, but how far we shall decentralize. What we seek, again, is a golden mean: we want to find the proper level in the organization hierarchy—neither too high nor too low—for each important class of decisions.

Over the past twenty or more years there has been a movement toward decentralization in large American business organizations. This movement has probably been a sound development, but it does *not* signify that more decentralization is at all times and under all circumstances a good thing. It signifies that at a particular time in history, many American firms, which had experienced almost continuous long-term growth and diversification, discovered that they could operate more effectively if they brought together all the activities relating to individual products or groups of similar products and decentralized a great deal of decision making to the departments handling these products or product groups. At the very time this process was taking place there were many cross-currents of centralization in the same companies—centralization, for example, of industrial relations activities.

There is no contradiction here. Different decisions need to be made in different organizational locations, and the best location for a class of decisions may change as circumstances change.

There are usually two pressures toward greater decentralization in a business organization. First, it may help bring the profit motive to bear on a large group of executives by allowing profit goals to be established for individual subdivisions of the company. Second, it may simplify the decision-making process by separating out groups of related activities—production, engineering, marketing, and finance for particular products—and allowing decisions to be taken on these matters within the relevant organizational subdivisions. Advantages can be realized in either of these ways only if the units to which decision is delegated are natural subdivisions—if, in fact, the actions taken in one of them do not affect in too much detail or too strongly what happens in the others. Hierarchy always implies intrinsically some measure of decentralization. It always involves a balancing of the cost savings through direct local action against the losses through ignoring indirect consequences for the whole organization. . . .

The automation of important parts of business data-processing and decision-making activity, and the trend toward a much higher degree of structuring and programming of even the nonautomated part will radically alter the balance of advantage between centralization and decentralization. The main issue is not the economies of scale—not the question of whether a given data-processing job can better be done by one large computer at a central location or a number of smaller ones, geographically or departmentally decentralized. Rather, the main issue is how we shall take advantage of the greater analytic capacity, the larger ability to take into account the interrelations of things, that the new developments in decision making give us. A second issue is how we shall deal with the technological fact that the processing of information within a coordinated computing system is orders of magnitude faster than the input-output rates at which we can communicate from one such system to another, particularly where human links are involved.

Let us consider the first issue: the capacity of the decision-making system to handle intricate interrelations in a complex system. In many factories today, the extent to which the schedules of one department are coordinated in detail with the schedules of a second department, consuming, say, part of the output of the first, is limited by the computational complexity of the scheduling problem. Often the best we can do is to set up a reasonable scheduling scheme for each department and put a sizeable buffer inventory of semi-finished product between them to prevent fluctuations in the operation of the first from interfering with the operation of the second. We accept the cost of holding the inventory to avoid the cost of taking account of detailed scheduling interactions.

We pay large inventory costs, also, to permit factory and sales managements to make decisions in semi-independence of each other. The factory often stocks finished products so that it can deliver on demand to sales warehouses; the warehouses stock the same product so that the factory will have time to manufacture a new batch after an order is placed. Often, too, manufacturing and sales departments make their decisions on the basis of independent forecasts of orders.

With the development of operations research techniques for determining optimal production rates and inventory levels, and with the development of the technical means to maintain and adjust the data that are required, large savings are attainable through inventory reductions and the smoothing of production operations, but at the cost of centralizing to a greater extent than in the past the factory scheduling and warehouse ordering decisions. Since the source of the savings is in the coordination of the decisions, centralization is unavoidable if the savings are to be secured.

The mismatch—unlikely to be removed in the near future—between the kinds of records that humans produce readily and read readily and the kinds that automatic devices produce and read readily is a second technological factor pushing in the direction of centralization. Since processing steps in an automated data-processing system are executed in a thousandth or even millionth of a second, the whole system must be organized on a flow basis with infrequent intervention from outside. Intervention will take more and more the form of designing the system itself—programming—and less and less the form of participating in its minute-by-minute operation. Moreover, the parts of the system must mesh. Hence, the design of decision-making and data-processing systems will tend to be a relatively centralized function. It will be a little like ship design. There is no use in one group of experts producing the design for the hull, another the design for the power plant, a third the plans for the passenger quarters, and so on, unless great pains are taken at each step to see that all these parts will fit into a seaworthy ship.

It may be objected that the question of motivation has been overlooked in this whole discussion. If decision making is centralized how can the middle-level executive be induced to work hard and effectively? First, we should observe that the principle of decentralized profit-and-loss accounting has never been carried much below the level of product-group departments and cannot, in fact, be applied successfully to fragmented segments of highly interdependent activities. Second, we may question whether the conditions under which middle-management has in the past exercised its decision-making prerogatives were actually good conditions from a motivational standpoint.

Most existing decentralized organization structures have at least three weaknesses in motivating middle-management executives effectively. First, they encourage the formation of and loyalty to subgoals that are only partly parallel with the goals of the organization. Second, they require so much nonprogrammed problem solving in a setting of confusion that they do not provide the satisfactions which are valued by the true professional. Third, they realize none of the advantages, which by hindsight we find we have often gained in factory automation, of substituting machine-paced (or better, system-paced) for man-paced operation of the system. . . .[6]

We can summarize the present discussion by saying that the new developments in decision making will tend to induce more centralization in decision-making activities at middle-management levels. . . .

The task of mddle managers today is very much taken up with pace setting, with work pushing, and with expediting. As the automation and rationalization of the decision-making process progress, these aspects of the managerial job are likely to recede in importance.

If a couple of terms are desired to characterize the direction of change we may expect in the manager's job, I would propose rationalization and impersonalization. In terms of subjective feel the manager will find himself dealing more than in the past with a well-structured system whose problems have to be diagnosed and corrected objectively and analytically, and less with unpredictable and sometimes recalcitrant people who have to be persuaded, prodded, rewarded, and cajoled. For some managers, important satisfactions derived in the past from interpersonal relations with others will be lost. For other managers, important satisfactions from a feeling of the adequacy of professional skills will be gained.

My guess, and it is only a guess, is that the gains in satisfaction from the change will overbalance the losses. I have two reasons for making this guess: first, because this seems to be the general experience in factory automation as it affects supervisors and managers; second, because the kinds of interpersonal relations called for in the new environment seem to me generally less frustrating and more wholesome than many of those we encounter in present-day supervisory relations. Man does not generally work well with his fellow man in relations saturated with authority and dependence, with control and subordination, even though these have been the predominant human relations in the past. He works much better when he is teamed with his fellow man in coping with an objective, understandable, external environment. That will be more and more his situation as the new techniques of decision making come into wide use.

6 The general decline in the use of piece-rates is associated with the gradual spread of machine-paced operations through the factory with the advance of automation. In evaluating the human consequences of this development, we should not accept uncritically the common stereotypes that were incorporated so effectively in Charlie Chaplin's *Modern Times*. Frederick Taylor's sophisticated understanding of the relations between incentives and pace, expressed, for example, in his story of the pig-iron handler, is worth pondering.

White Collar Workers

ALBERT A. BLUM

Michigan State University

I. INTRODUCTION

Machines alter or replace labor. These are among their major purposes whether a machine is part of an automobile plant's assembly line or a bank's check-sorting system. Thus, it may be that a check-sorting machine or a computer may or may not result in immediate job loss, but it surely will at least result in changes in the jobs worked. The employee who did the replaced job will have to use different skills whether he works at the new machine or is transferred to another. (Even if changed to a similar job, it will be at a different place—with different implications for him and the work organization.) If the machine is completely new, doing work not previously done, a person employed at that machine would require skills new in the company, with all of the concomitant labor relations problems which follow as a result.

Since machines alter work, there should be some changes in skills, salaries paid, and conditions of work. Since some machines replace workers, there should be some changes in employment. These changes should have some implications for education, training, and unions. And yet, when one explores the research so far done, one notes a somewhat confusing pattern— namely, a lack of pattern. Are there more or less opportunities for promotions as a result of automation? It depends, report the researchers. Does automation raise or lower skills? Again the answer is that it depends. And on what does it depend? Frequently, it depends upon the research read, for even given two solid and substantial research studies, nontheless, the findings may differ.

Reprinted from Albert A. Blum, "Computers and Clerical Workers," Document D 1–68 of the Third International Conference on Rationalization, Automation and Technological Change, sponsored by the Metalworkers' Industrial Union of the Federal Republic of Germany (Oberhausen, Germany, 1968), by permission of the author.

Why is this so? I would suggest that one of the reasons for this is that such conferences as this one, added to affluence, have confused (with good and bad results depending upon one's values) the effects which automation might have had in the absence of such factors.

One of the problems researchers attempt to control for is the Hawthorne effect—namely, that people know they are being studied and consequently alter their behavior. But is there not a broader Hawthorne effect at work for which researchers cannot control? In the last decade or so, conferences, committees, scholars, unions, companies, governments, international organizations, and so forth have been discussing the implications of automation. It is no wonder that such activities have had an overall Hawthorne effect in that management policies in this field have been influenced by such discussions.

An executive in the nineteenth century, as he installed a new piece of technology, would have laid off his workers, perhaps unhappily, but with little thought of what society, his peers, or a union would think of him. But today, a manager is concerned when he installs a computer that if he discharges workers, first, these discharges would lower the morale of the rest of his office employees; second, a more socially worthwhile procedure would be to let attrition, transfers, or retraining take care of the surplus of employees; and third, rather than reduce salaries, one should pay the employee his old rate for a period of time. And who has convinced management to take these more expensive and perhaps less technically efficient steps? His government, his management association, his peers, and his union have so convinced him and affluence has permitted him to perform these acts. Once the affluence disappears, however, will the implications of automation continue to be cushioned? That may be for an unhappy tomorrow to answer but as of today, as we begin to explore the effect of computers on clerical workers, we have to remember that there may be a backlog of problems which a recession may let loose. Thus time has to be taken into account in studying automation's implication.

That management recognizes the potential, more than the immediate problems resulting from automation is reflected in one survey of management attitude. Of the 80 companies which had recently installed automated equipment, only two reported any increase in interest among their employees in unionism. And yet in this same survey, 25 percent of the executives think that in the long run, their clerks fear automation while more than one third think that in the long run, automation will result in their clerks joining unions.[1]

Another aspect of this short-run versus long-run effect is whether automation results in a displacement of clerical employees. Although this is not a major theme of this paper, automation generally appears not to have resulted in immediate discharges of clerks. What frequently happens is a decline in the number of new empolyees hired over the long run while

[1] Albert A. Blum, *Management and the White Collar Union,* American Management Association, 1963, p. 66.

attrition and transfers are the immediate techniques used to reduce staff to the size desired. In addition to permitting a company to carry the costs involved in such a practice, affluence has also helped increase the demand for paper so that even with the increased productivity of the machines, the demand for the products of those machines has also gone up, thereby lessening the need to discharge clerks. Moreover, the time involved in installing automated equipment gives management an opportunity to make the adjustments needed. These factors are among those which help explain why computers do not cause major immediate displacements. But our concern in this paper is not so much with the impact of automation on those no longer at work but on those who remain employed.

II. EFFECTS OF AUTOMATION

Professor Kenneth Walker comments in his international comparison of the impact of automation on non-manual workers that "it is in the group of 'clerical workers' [more than other groups of nonmanual workers] that the most striking impact of automation and advanced technology has so far been apparent. . . . Operators of office machines, bookkeepers, and other workers maintaining financial and other records have been the most radically affected. Stock control, invoicing and billing, production control and payroll work—including maintenance of associated personnel records— are other activities which have been computerised."[2]

To analyse this impact, we will first examine the job and skill structure in an office: in the wake of the computer, which jobs disappear, which jobs increase, has the computer raised or lowered clerical skills? Second, we will see how clerks move from one job to another: has the computer resulted in increased or decreased promotion possibilities? Third, we will analyse the wage structure: has the computer lowered or raised salaries? Fourth, we will examine working conditions: has the computer affected the hours worked by clerks? And, fifth, we will briefly examine the overall implications of these changes.

Changes in Job Structure and Skill Mix

Although there are those who disagree, most observers claim that the computer coming into the office has had the following effects: the lower level, routine jobs have been reduced in number while the higher level, more skilled clerical jobs have increased in number. The Bureau of Labor Statistics in the United States, for example, studied a number of white collar occupations with the following results:

Communications Industry. Clerical employment is being reduced [while] professional and semi-professional employment, engineers, programers, analysts, will continue to grow.

2 Kenneth A. Walker, "Automation and Nonmanual Workers," *Labour and Automation Bulletin* No. 5, Geneva: International Labour Office, 1967, p. 10.

Banking Industry. Employment is expected to continue to increase but at a decreasing rate.... Bookkeepers are hardest hit by automation.... For example, at one multibranch bank, within 18 months after the start of conversion to electronic data processing (EDP), the bookkeeping staff of 600 had been reduced to 150, and the data processing staff had grown to 122, a net reduction of 55 percent.... The number of supervisory personnel has been growing.... Introduction of automation has modified old jobs and created new ones.... The principal new occupations are EDP equipment operator, programer, systems analyst, encoder, and EDP clerk.

Insurance Carriers. Insurance carrier employment will continue to grow slowly in the next five years.... Office workers, who represent about three-fourths of all insurance employees, may grow at a slower rate because the number of employees in clerical occupations will not increase at all.... Entry clerical jobs for girls will not be as numerous as formerly [for]...field office record-keeping is being eliminated.... EDP programing, systems analysis and operating jobs will continue to increase.... Keypunch operator jobs probably will not increase between 1965 and 1970 [because new equipment is making this latter skill obsolete].

Federal Government. Employment is expected to continue to rise but only moderately.... Routine clerical jobs, including keypunch-operator positions, are expected to decrease significantly.... Accounting and statistical clerical occupations, for example, may decline by 14 percent between mid-1963 and mid-1968. ...Professional and technical workers are expected to increase significantly.[3]

A Canadian scholar reports that "a survey of the research literature indicates that the introduction of electronic data processing will create widespread displacement among those engaged in routine clerical operations in the department and units affected by the change. Apparently the extent of this dislocation may involve from one third to 75 percent or more of the clerical jobs, depending upon the nature and extent of the computer applications."[4]

A recent ILO study echoes this point of view. It first cites an ILO report in 1959 that "the introduction of electronic computers in offices has...created a need for more highly skilled technical personnel, at the same time as it has done away with the need for large numbers of semi-skilled workers on routine jobs." The author of the ILO study then concludes that research since 1959 supports this view.[5]

A recent British study indicates that automation has resulted in the largest loss of jobs among clerks and machine operators on non-electronic data processing equipment with some increase among system analysts, and programers with most of those replaced being female employees.[6]

[3] U.S. Department of Labor, *Bureau of Labor Statistics Bulletin* No. 1474, "Technological Trends in Major American Industries", Washington, February 1966, pp. 222–258.

[4] J.C. McDonald, *Impact and Implication of Office Automation,* Department of Labour, Ottawa, May 1964, p. 18.

[5] Walker, *op. cit.,* p. 13.

[6] Manpower Research Unit, Fourth Report, "Computers in Offices," included in Organisation for Economic Cooperation and Development (OECD), *Manpower Aspects of Automation and Technical Change,* Zurich, 1966, pp. 211–215.

Basically what these writers and others are arguing is that automation raises the overall skill level in the office because the bottom of the hierarchy of skills is reduced while the peak is increased and eventually one may see an inverted pyramid of skills in the office.[7] But there are those who argue against this assessment. Thus, Ida Hoos, in her studies of automation in offices on the United States west coast, states that automation does displace lower level bookkeepers, filing, and ledger clerks, but she also argues that there has been no marked increase in the number of highly skilled jobs and that whatever increase has occurred could be found in such relatively unskilled employment like key-punch operators. She also believed that for every five jobs eliminated by automation, only one new one has been created.[8]

Another author concludes from his reading of the research that "the prediction that office automation will result in more interesting, less tedious work for the clerical work force seems only barely supported by current research." He argues that office, like production, automation can perpetuate "routine, uninteresting jobs." Thus he points out that a "clerical worker who is responsible for several routine office tasks is probably experiencing the greatest amount of diversity she will ever encounter. The introduction of automatic data-processing equipment will probably find her transferred to a key punch machine, card collator, or card sorter, where work routines become more formal, yet opportunities for mobility about the office is narrowed".[9] Thus, the job content of clerical jobs, many observers report, is narrowing with numerous exceptions, however.

How can one explain the differences in the findings—namely, one group of researchers finding a net increase in higher level jobs, and the other finding no increase—in fact, at times, a decrease in the level of skills. One explanation is time. Sola and Hoos, for example, tend to refer to changes caused by the earlier stages of automation at the end of the 1950's. Then, it was true that a bookkeeper, rather than finding her employment ending as did her job, might be transferred to a position as punch card operator. As one French observer noted: "Automation nearly everywhere has been

---

[7] See also Edgar Weinberg, "Experiences With the Introduction of Office Automation," *White Collar Report* May 9, 1960, pp. C1-C5; Roy Helfgott, "EDP and the Office Work Force," *Industrial and Labor Relations Review,* Vol. 19, July 1966, pp. 508–510; George E. Delehanty, "Office Automation and the Occupation Structure," *Industrial Management Review,* Spring 1966, Vol. 7, No. 2, pp. 99–109; Trades Union Congress *Conference Report 1965,* "Non-Manual Workers," pp. 5–6; August L. Cibavich, "An Insurance Company Automates," *Employment Security Review,* July 1962, pp. 40–41.

[8] Ida Hoos, "Computers in the Office," *Harvard Business Review,* July-August 1960, pp. 103–106 and her *Automation in the Office,* Washington, Public Affairs Press, 1961, pp. 23–58. See also Allen I. Kraut, "How EDP is Affecting Workers and Organizations," *Personnel,* July 1962, p. 43.

[9] Frank C. Sola, "Personnel Administration and Office Automation: A Review of Empirical Research" *ILR Research,* Vol. VIII, No. 3 ,1962, pp. 4–5; Floyd Mann and Lawrence W. Williams, "Observations on the Dynamics of a Change to Electronic Data Processing Equipment," *Administrative Science Quarterly,* September 1960, p. 253.

a very gradual process. One stage consisted of specialized punched card equipment, while the advent of computers was reflected first in the operations previously separated in the punched card department.... The change has therefore been slight as far as office staff are concerned."[10] Claudine Marenco, based on studies in France, also makes this worthwhile distinction between pre- and post-automated phases and the kinds of jobs which remain after the technological change.[11]

Thus, what may be occurring is that as automation progresses, such relatively routine jobs as punch card operators will join other clerical jobs in disappearing, as the United States Department of Labor, cited earlier, indicated.

Two changes, then, appear to be happening, in the long run, to the job and skill mix in the automated office. One is that in some automated offices there will be a greater proportion of highly skilled clerical employees. But the cautionary note just mentioned must be reiterated. To use an argument similar to one used by two California professors when they examined the overall impact of automation on skills, let us assume that before the installation of a computer, an office unit had 200 employees, 150 doing routine clerical work and 50 doing highly skilled work. Thus, there was a 3:1 mix of unskilled to skilled. After the computer was installed, there remained only 100 workers, with 60 being unskilled and 40 skilled. The mix had dropped to 6:4—clearly a raise in the skill mix in the office but at the cost of 90 unskilled jobs and 10 skilled jobs. Even if the number of employees remained the same but output went up markedly (thereby reducing the number of employees who might have been hired), and the skilled and unskilled proportion changes to a 1:1 proportion, the result is an overall rise in skill levels with constituent marked implications for education, training, transfers, and employment.[12]

Another long-term answer is offered by W. H. Scott in his OECD study of office automation. He discusses the future structure of the office:[13]

> It seems clear that computer systems are accentuating certain established trends in the composition of the clerical and administrative work force. During recent decades, as the size of this group has grown, clerical work has been increasingly simplified, routinised and mechanised. Thus a growing proportion of employees are on semi-skilled work, operating machines or performing

10 Jacques Urboy, "A Tentative Interpretation of a Number of Case Studies of Firms and Industries Using Office Computers," in OECD, *Manpower Aspects...op. cit.,* p. 180.

11 Claudine Marenco, "The Effects of the Rationalization of Clerical Work on the Attitudes and Behavior of Employees," in Jack Stieber, ed. *Employment Problems of Automation and Advanced Technology: An International Perspective,* New York, St. Martin Press, 1966, pp. 412–429.

12 Paul Sultan and Paul Prasow, "The Skill Impact of Automation," in Subcommittee on Employment and Manpower, Committee on Labor and Public Welfare, U.S. Senate, *Exploring the Dimensions of the Manpower Revolution,* Vol. 1, Washington, 1964, p. 552.

13 W. H. Scott, *Office Automation: Administrative and Human Problems,* OECD, Paris, 1965, p. 95.

routine clerical tasks, and most of this category are now women. Our studies suggest...that computers are accelerating this trend. Admittedly there is the new skilled element of computer supervisors, programmers and operators, but their number is relatively small and,...their advent does not by any means offset the decline in the proportion of skilled clerks in other sections. The future trend is somewhat uncertain. On the one hand, as we have said, improved input procedures may reduce the proportion of routine clerks, and it may be that more integrated computer systems will call for rather more skilled personnel in connection with the interpretation and application of "output." However, it is probable that the computer supervisor and programmer will assume an ever more important role in the future as systems become more complex, and that their function and status will become managerial. In this event, in the fully automated office, the clerical group may consist overwhelmingly of routine workers.

What Scott is predicting is that probably many of the highly skilled jobs resulting from automation will move out of the complex of clerical jobs—perhaps into management or into the ranks of professional or near-professional levels leaving clerks who remain in the office involved in machine-oriented, routinised work, with a sharp gap between their level of work and those above them. And some union leaders believe that if this be the case then unionism will come easier to the nonmanual worker for "when the white collar person becomes a baby-sitter for an automated machine, pride of work gets drained out of his job, and he is going to try to join with his fellows."[14]

Job Changes Through Transfers and Promotions

"The feeling that as an individual he cannot get ahead in his work," says C. Wright Mills, the late famous sociologist, "is the job fact that predisposes the white-collar employee to go pro-union.[15] Another factor which may predispose him to move into a union would be the inability, once his job disappears because of a computer, to be able to transfer to another. And it is to these different, but still related topics, which we now turn.

A recent ILO survey comments that an earlier ILO 1959 report on the impact of automation on promotions in offices appears to have been wrong. The 1959 report concluded that with automation in offices:

> A few workers will have more interesting and better-paid jobs, but many more will remain at the machine-operating levels, where chances of advancement are small, and others will be transferred to other departments to do work of a nature equivalent to that which they have previously been doing.

In 1967, the author of the later ILO report concludes that "in general, the published studies indicate that as yet the net reduction in promotion

---

[14] Jack Barbash, cited in *White Collar Report,* January 2, 1961, pp. A-1, A-2; Blum, *op. cit.,* pp. 59–71; Walker, *op. cit., passim.*
[15] C. Wright Mills, *White Collar,* New York, Galaxy, p. 307.

opportunities as a result of office automation has not been significant in relation to office employment as a whole." Moreover, in his survey of unions and employers, he further reports that "both employers and trade unions supplying information...considered that the net effect of automation in clerical work had been to improve the prospects of promotion."[16] Could it be possible that both ILO reports are correct and it is just that there have been some safety valves which have permitted the adjustment processes to automation to work relatively well for the present, at least?

For purposes of analysis, one can group office employees in three categories as does Frank C. Sola: "First, those who do routine office work involving typing, filing, sorting, and so forth. Second, personnel with both experience and seniority whose work requires slightly more knowledge of company operations. Third, those who hold managerial positions of either a technical or an administrative nature."[17] Let us examine what is happening to these groups as computers move into their offices.

First, automation appears to be cutting off the middle step in the old promotion ladder in that the second group appears to be generally growing smaller. There is even some evidence that some of the third group is disappearing but a member of the third group is relatively easy to transfer. He can often take over administrative duties; he frequently has the kind of technical skills which permit relatively easy transition, through training, to the technical positions at the computer centers. But the second group remains difficult to retrain and transfer—particularly if he is of advanced age.[18]

Second, those with less skills (those in the first group mentioned earlier) can be transferred to similar routine jobs at the computing center or to other jobs in the company. Those who cannot move to another job will probably be kept on and since so many of these employees are women, attrition will eventually reduce the overload.

Two of the safety valves have thus been mentioned—women and older workers. If there is only a limited number of promotional opportunities available, companies tend to favor the young male for the higher level positions when they do not look outside for new employees, and they permit time, through attrition, to take care of the women and older workers. Management frequently feels that patience will be the female and the older clerk's reward—namely, marriage and/or pregnancy for the girl and retirement, perhaps at an earlier age, for the older male. Thus, time, women, and age provide the safety valves which permit the job progression ladder to remain somewhat steady in the face of an automation explosion—particularly if one is a young, male clerk.[19]

[16] Walker, *op. cit.,* pp. 20–23.
[17] Sola, *op. cit.,* pp. 5–6.
[18] Ida Hoos, "The Impact of Office Automation on Workers," *International Labour Review,* Vol. 82 (4), October 1960, pp. 370–371.
[19] McDonald, *op. cit.,* pp. 21–22; *Labour,* February 1963, p. 34; Helfgott, *op. cit.,* p. 510.

But over time, however, the promotional ladder is likely to develop some missing steps. Thus, a number of studies indicate that there is a reduction in the number of middle-level white collar positions—caused perhaps less from automation directly but from the re-organization of the office resulting from automation such as the centralization of the work. Moreover, as mentioned earlier, there may be an increase in the number of higher level jobs which will be frequently of a technical nature.[20] If these trends continue over time, it will be more and more difficult for a clerk to rise in the hierarchy of a company. The middle steps will be missing. In fact, many white collar employees have already found their upward mobility blocked by these missing steps.[21]

It is this blocked mobility, particularly among the men in dead-end jobs, which may make these men members of the "clerical proletariat" and prompt them into moving into unions.[22]

Salary Structure

There has not been much direct study of the effects automation has on salaries in the office but what studies there have been seem to indicate the following tendencies. First, there have not been many increases in salaries directly resulting from the computers. Thus, all levels of clerks continue to receive about the same rates of pay after automation as before—with only slight modifications. This has caused some distress for those clerks who thought automation would result in raises. Second, those employees whose skills have been downgraded have often continued to receive their former salaries, but their jobs have been "red-circled"—that is, the job itself calls for a lower salary but as long as that individual continues to work at it, his salary is protected. (Sometimes there is a limit on the length of time of the protection.) This method of keeping up the morale and income of affected employees sometimes has negative effects upon the morale of new employees who are working at more responsible positions but at no, or little, difference in salary as compared with the older employees at work at more routine tasks. On the other hand, a number of companies have let the salaries of the transferred, senior workers go down over a period of

---

[20] Walker, *op. cit.,* p. 21.

[21] See articles cited in Albert A. Blum and Gil Shaal, "Automation and Plant Level Issues: A Survey of the Literature," in a forthcoming study of automation and plant level issues to be published by the Automation Unit of the International Labour Office; "Effects of Mechanization and Automation in Offices: II," *International Labour Review,* Vol. 81 (3), March 1960, pp. 266–270; J. Siegman & B. Karsh, "Some Organizational Correlates of White Collar Automation," *Sociological Inquiry,* Winter 1962, p. 114–115; Albert Kushner, "People and Computers," *Personnel,* January-February 1963, pp. 33–34; Marenco in Stieber, *op. cit.,* p. 417; Claudine Marenco "Psychological Incidences of Office Work Rationalization on Employee States," *Trade Union Information,* No. 35.

[22] See also J. R. Dale, *The Clerk in Industry,* Liverpool, Liverpool University Press, 1962, p. 88.

time—if not immediately—with frequent morale problems. Third, the new jobs which have resulted from automation have frequently not been placed in the salary structure at very impressive positions. Fourth, it is true that in many offices the overall effect of automation has been to raise the salary level but this has not been so much because individual jobs have had their salaries raised but rather because the overall skill mix in the office has changed. In those offices where the computer results in there being proportionally fewer unskilled clerks and more skilled clerks, then, the over-all salary level in the office will have been raised. Fifth, since automation has resulted in many of the lower-level jobs being more oriented toward machines and as a result can be measured more accurately, job evaluation schemes are becoming more appropriate for clerical work with production standards being set and efficiency experts flocking to offices, figuring out ways of increasing output.[23]

Of course, salaries are not determined in a vacuum. Many companies have markedly raised salaries after the installation of the computer but often as the result of a tight labor supply rather than automation. Moreover, a general tendency has been at work among office employees for most of the period since the Second World War—at least in the United States and a number of other countries—namely, that manual workers have seen their wages and supplementary benefits increase at a more rapid rate than those of the nonmanual employees, thus narrowing the compensation gap between both groups. Automation has resulted in no particular break in this tendency. As mentioned earlier, it has in fact caused some salary situations with which clerks are dissatisfied. Moreover, the computer may eventually cause differentials to widen between the upper and lower skill levels of nonmanual employees with very few clerks in the middle to bridge these differentials. All these factors may give unions an opportunity to focus on salary inequities as a reason to join unions or if the clerks are already in unions, give unions problems which they might try to solve.[24]

Impact on Working Conditions

There are many areas of changed working conditions resulting from the computer in the office which could be discussed here but lack of space and time permits the discussion of only one—namely, the increase in shift work. That there has been an increase in the number of clerks who have to work on shifts seems to be reflected in most of the research done on the topic. Some clerks, particularly young male employees, welcome the opportunity to work afternoon and night shifts in order to earn more money. Some

---

23 Sola, *op. cit.*, pp. 8–9; Helfgott, *op. cit.*, pp. 509–510; Walker, *op. cit.*, pp. 14–17; "Effects of Mechanization...II," *op. cit.*, pp. 270–273; Scott, *op. cit.*, p. 96.
24 Sola, *op. cit.*, p. 9, Blum, *op. cit.*, pp. 29–47; Adolf Sturmthal, ed., *White Collar Trade Unions*, University of Illinois Press, 1966, *passim* & p. 394–397.

women like the afternoon shift for it fits into their domestic life while other women do not want to work late shifts at all. Some older clerks, it is reported, do not like shift work either.[25]

Thus there is no clear pattern to the reaction of clerks to shift work. Very few of the studies, however, focus on what these reactions mean to clerical unions. Unions could help satisfy the clerks' wishes when automation requires that computers be used for more than eight hours a day, five days a week. They could demand that seniority plus willingness to work shifts be the criteria determining who should work shifts. This has already been true in certain offices where the unions' demand for seniority in determining who should work shifts has been one of the major attractions the union has had for the women at work at nonmanual jobs.[26]

III. CONCLUSIONS

In a world which is becoming more and more nonmanual at the same time that it is becoming more and more automated, the implications of automation on nonmanual employees become increasingly important—not only for the clerk already at work but also for the young girl, now at school, who aspires to a nonmanual position. Some of these implications have already been pointed out—namely, the fact that time and social consciousness have cushioned the impact on the already employed clerk whose job has been taken over or altered by a computer.

But the gap between the less skilled and the higher skilled clerk appears to be growing larger and larger. Some companies have tried to bridge the gap through training programs. Frequently, much of this training, particularly for the jobs at the computer center, has been offered by the company making the computer. But, how should the person be chosen from among company employees to train for these higher level jobs? Unions have often demanded that seniority be the main criterion; management has favored skill, frequently determined by objective tests which normally benefit younger workers more attuned to such examinations. Moreover, companies favor sending younger men, not older men or women, to those training programs because of the long-term benefits they hope will result from the trained youth. And frequently, rather than go through the trouble of developing skills among their own staff, companies will hire the graduates of these or other training programs regardless of their former employment or lack thereof. When these trainees (be they old or new employees), come to work at the office, there will usually be an additional

---

[25] Walker, *op. cit.,* pp. 17–19; Blum and Shaal, *op. cit.,* section on hours of work; U.S. Department of Labor, Technological Trends..., *op. cit., passim.*
[26] Norman Bozeman, "Unionization of White Collar Workers (Clerical) on the Communications Industries," Unpublished manuscript, University of Michigan, 1962, cited in Blum, *op. cit.,* pp. 56–58.

period of on-the-job training on the new computer. On-the-job training also is the normal technique used to train those clerks who transfer from routine jobs to the other relatively routine jobs at the computer center.[27]

Because of a declining rate in growth of employment in many offices and because of the raising of the skill level in some of the offices, many personnel managers are demanding that new clerks be first employed at a higher level of skill. In the United States, secondary school graduation and even some college experience are becoming the prerequisites for employment at clerical jobs. This, in itself, has many implications for our educational system. Moreover, automation makes it more and more necessary that society not only provide our young with more education but with the kind of education to make them more flexible in the market. Thus, schools must ready our young for change; and so must the companies that hire them. The latter cannot wait until technology is upon them before instituting their own training programs or in working with outside adult education programs. It must involve itself and its employees in these programs continuously so that they will be ready for the kinds of training programs which will prepare them for the computer. Many studies indicate that, given two employees, the one who has had frequent training experiences since his school days will be more likely to attend another training program than will the employee who has not had any such experiences since his pre-employment days.

But the school's and society's responsibility are equally great. Since it is clear that many companies will not spend the money and take the chance to retrain employees who do not have adequate basic training, schools must provide it. And the kinds of skills required are not so much the specific ones required for specific jobs, but rather broad basic skills which will permit the clerk to transfer to a new job with a minimum of additional training. Thus, there is a need for the meshing together of the school's educational program with the on-the-job training programs of industry and the overall adult education programs prevalent in society and not to treat them as separate entities as has frequently been the case.[28]

27 B. G. Stacey, "Automation and the Office Worker," *Labour Monthly,* February 1965, pp. 79–80; Marvin J. Levine, "The Impact of Automation Upon Clerical Employment," *Personnel Administrator,* July-August 1965, pp. 33–34; "Effects of Mechanization. . . . , *op. cit.,* pp. 260–264; Sola, *op. cit.,* pp. 7–8; Ida Hoos, "Technology, Retraining and the Training Director," *Business Topics,* Winter 1966, pp. 48–50; *Proceedings* of First Annual White Collar Worker Retraining Workshop, Michigan State University, May 1964, reproduced by Bureau of National Affairs, *White Collar Report,* May 1965; Kraut, *op. cit.,* p. 49. For a discussion of the role education and training can play for clerks, as well as other related topics, see *Report* of the Director General, International Labour Conference *51* Session, "Part I: Non-Manual Workers: Problems and Prospects," ILO, 1967, pp. 36–82 and *passim.*

28 For overall treatment of these problems, see Albert A. Blum (coordinating editor), "Manpower Adjustment Programmes" (nine different countries), *Labour and Automation Bulletins* 4, 6, and 7, International Labour Office, 1967 and 1968, Chapters 4 and 5 in each volume and Albert A. Blum, "Job Skills for Automated Industry," *Management of Personnel Quarterly,* Winter 1964, pp. 24–31.

The Future Computer Utility

PAUL BARAN

Institute for the Future

The most important policy issue regarding the use of computers in the next decade will involve the creation of a "national computer public utility system." It may seem strange to think of the computer in terms of a public utility, since ostensibly a computer is a machine that a customer buys or rents for his own use. But the recent emergence (only in the last year or two) of the possibility of "time-sharing"—whereby thousands of individual terminals, located in homes or offices, can be hooked into giant central computers through the use of telephone lines and used for information-gathering, ordering and billing services, etc.—makes the question of such a utility system anything but academic.

In Great Britain, a bill has been introduced into Parliament to allow the General Post Office (G.P.O.) to sell data-processing services and facilities. The G.P.O. in Britain has a monopoly of the telephone and telegraph services; and it is envisaged that it will also sell a network of computer services to various customers on an *ad hoc* (e.g., a dime for a telephone call) or contract basis. In the United States, of course, we do not have a single government system. Most of the wire services are handled through A.T.&T., although the General Telephone Company and hundreds of smaller independent companies are hooked into the national system. So we have various alternatives. We can create a new national monopoly-utility. We can license a competing national system to avoid extending the present scope of A.T.&T. (just as COMSAT was set up as a new international communications system). Or we can permit a system of laissez-faire.

The problem, in short, is how do we "hook up" the unregulated purveyors of computer data-processing services such as General Electric with

Reprinted from Paul Baran, "The Future Computer Utility," *The Public Interest*, 8 (Summer 1967), pp. 75–87, © by National Affairs, Inc. Used with the permission of National Affairs, Inc. and The RAND Corporation.

the regulated communication companies, such as A.T.&T. and Western Union. This article seeks to explore the alternatives.

THE COMPUTER REVOLUTION

To understand the problem of a national computer information system, we might look at the historical development of earlier transportation and communication systems. Our first railroads in the 1830's were short routes connecting local population centers. No one sat down and laid out a master plan for a railroad network. Over time, an increasing number of separate local systems were built. A network gradually grew as economic pressure caused new links to be built spanning gaps between individual routes. Similarly, we did not start to build a nationwide telegraph network in the late 1840's—only independent telegraph links; nevertheless we soon had an integrated nationwide network. Even the name, *Western Union,* recalls the pattern of independent links joined together to provide a more useful *system*. Nor was a nationwide telephone system planned in the 1890's. Yet today we have a highly integrated telephone network.

Such patterns of growth are not accidents. Communications and transportation are services that historically tend to form "natural monopolies." The reason is understandable. It is far cheaper to share the use of a large existing entity than to build one's own facilities.

Now the newly developing computer utilities have the same properties as communication and transportation networks, and the same inherent tendency to be self-agglomerating. It is cheaper to share information by tying together independent data-processing systems than by building a large number of duplicating systems without interconnection. Such "sharing" has been made possible by new technological developments.

These new developments in computer technology are of such significance as to affect materially the nature of our economic and social life. Numerical processing, the business that gave the computer its name, has already been relegated to secondary importance, as the new generations of computers are being used for such tasks as diagnosing illness, simulating business operations, creating elementary teaching machines, reaching routine business decisions on investment and inventory policy, scanning addresses of letters, and sorting mail. This open-ended list is but an indication of what tomorrow's computer will be doing. The word "computer" is by now a misnomer. It is the *automation of information flow* that is really at the heart of the computer revolution.

The speed and facility with which computers manipulate numbers is already an old story. The new computers process and rearrange words as readily as their predecessors handled numbers. Today, large files can be processed, stored into the memory banks of the computer, and rearranged upon command. Information is being retrieved in an increasingly abstract and sophisticated manner.

To take a simple and even "absurd" example. Suppose you wish to send off several slightly different, angry letters to editors of different periodicals. You would type a rough draft onto an electric typewriter device; add a few symbols to rearrange the orders of the paragraphs; call up some stock phrases which can be varied; add your own previously defined symbol to specify your list of addresses; and when satisfied, type a single word, "Go." Out of the typewriter, at 150 words per minute, would come a series of perfectly typed, "individual" letters. The publication deciding to publish your letter would hand it to a typist. She would type your letter into the computer, specifying a desired column width. Such details as decisions on hyphenation, or the insertion of blank spaces to neatly justify the right-hand margin, are provided by the computer. (The computer memory might even contain a dictionary to prevent most misspellings.) At some later date in the future, automatic print readers might even eliminate the need for the publisher's typist.

All this may sound expensive and impractical. We do not want to buy an entire computer for such simple tasks. It is like buying a cow for a little cream for our coffee. But with the advent of the computer utility, we need not do any such thing. We would merely use a small portion of the large capability possible in our new computer systems, which are becoming ever more powerful.

This major advance in spreading the access of the computer to an ever wider segment of society is known as "time-sharing." Our newer computers are so fast, relative to human response, that a single computer can now rapidly serve a large number of users and maintain the illusion that each has his own computer. The conversion of raw computer speed into simultaneous service to a large number of users is a key aspect of the coming computer revolution, for it spreads the benefits of computer technology widely. With a marked decline in the cost of computing power each year, the usable market for newer information-processing systems can only increase with time.

And while it may seem odd to think of wanting to pipe computing power into homes, it may not be as far-fetched as it sounds. We will speak to the computers of the future in a language simple to learn and easy to use. Our home computer console will be used to send and receive messages—like telegrams. We could check to see whether the local department store has the advertised sports shirt in stock in the desired color and size. We could ask when delivery would be guaranteed, if we ordered. The information would be up-to-the-minute and accurate. We could pay our bills and compute our taxes via the console. We would ask questions and receive answers from "information banks"—automated versions of today's libraries. We would obtain up-to-the-minute listing of all television and radio programs. We could use the computer to preserve and modify our Christmas lists. It could type out the names and addresses for our envelopes. We could store in birthdays. The computer could, itself, send a message to remind us of an impending anniversary and save us from the disastrous consequences of forgetfulness.

Within the next few years we can expect to see about 100 *separate* general-purpose, time-sharing computer systems. These early general-purpose systems can, for the most part, only serve customers economically within 50 to 100 miles from the central computer. *The economic limitation on market size is the high cost of long-distance telephone lines.* We may therefore expect to see many small, independent "computer utilities"—at first.

THE PROBLEM OF REGULATION

The computer field, while dominated by a single large company—IBM —has bred a host of small, fast-moving competitors that have kept the leader on its toes. Competition in the computer business is relatively fierce and the number of losers is large. But in long-distance communications, we have a history of protective governmental regulatory action which allows less room for free enterprise competition. The conflict begins as the totally unregulated computer companies rub up against their highly regulated communications suppliers. The technological growth of computers may be limited by the regulatory structure for data transmission unless new regulatory doctrine is created.

While the term "computer utility" is commonly used to describe the new computer-communications networks, these are markedly different from the classical utility. A utility is a "natural monopoly" in which the government issues an exclusive franchise to provide service to the public on a noncompetitive basis. In return for the monopoly granted, the utility acquires a set of obligations. This includes the requirements to provide service to all comers on an equal basis, including those that are highly profitable and those that are not. We pay the same for electricity whether we live next door to the electric generator or on the fringe of the city.

On the other hand, a potential user of a "computer utility" in a big city can buy his service from any one of perhaps a dozen or more strongly competing time-sharing systems. The user is not restricted to doing business with any one company. If one is not satisfied with the service, or is concerned about price, one can always "go" elsewhere. Similarly, any single computer installation is not forced to serve all potential customers on an equal basis. The big customer may expect preferential treatment, either in terms of prices charged or speed of service. (Time-sharing systems have their own "peak-load" problems.) But there are several reasons why some form of regulation may be required. Consider one of the more dramatic ones, that of privacy and freedom from tampering. Highly sensitive personal and important business information will be stored in many of the contemplated systems. Information will be exchanged over easy-to-tap telephone lines. At present, nothing more than trust—or, at best, a lack of technical sophistication—stands in the way of a would-be eavesdropper. All data flow over the lines of the commercial telephone system. Hanky-panky by an imaginative computer designer, operator, technician, communications worker, or programmer could have disastrous consequences. As time-shared computers come into wider use, and hold more sensitive information, these problems can only increase. Today we lack the mechanisms to insure adequate safeguards. Because of the difficulty in rebuilding complex systems to incorporate safeguards at a later date, it appears desirable to anticipate these problems.

Apart from the issue of potential threats to privacy some *de facto* regulation of time-shared computer systems is already appearing—but largely as a by-product of the Federal regulatory rules established for telephone and telegraph communications of an earlier era; and this is not wholly satisfactory for the problem of the computer utility.

The utility regulatory process has been fundamentally a backward-looking affair. Rules under which one must live tomorrow have been based on experience of the past; regulatory bodies rarely lay out price tariffs on the expectation of new technology to come. In at least one major instance, "forward-looking tariffs" (tariffs estimated on future costs) have been specifically rejected by the FCC. Based on existing regulations, any communications tariff philosophy for time-shared computers would be a legacy from a different era and for a communications system designed for a different purpose.

A new tack was taken in November, 1966, when the FCC, to the surprise of the computer industry, announced that it planned to consider its own role in the domain of the "computer-utility." An inquiry of such magnitude usually takes several years before being resolved. But it is clear that the FCC appreciates that there are new issues that need attention. For it is increasingly evident that the new generation of time-sharing systems is best served by a somewhat different communication—one specifically designed to operate between computers, and from computers to people rather than the traditional voice telephone. The common-carrier voice communications systems of today were designed primarily to provide a voice-to-ear or type-writer-to-typewriter link between humans. Today's communications regulatory doctrine still regards the computer merely as another user of these existing telephone, typewriter, and telegraph networks. But this is the jamming of a size ten foot into a size five shoe. The communications characteristics we desire in computers are so fundamentally different from previous systems that any carry-over from the past would be procrustean. For example, the computer user wants short, high-speed spurts of data. He is intolerant of the errors caused by background noise. He doesn't want to have to pay for a long-time telephone-call-like connection, for his line will be idle most of the time. He would like to share it with other similar, short-burst, computer "party-line" users serving other systems—with time sliced very finely. The subtle nature of the new and more complex technological interface between the older telephone communications system and the computer is still not completely appreciated, even by the computer industry.

THE PLAYERS IN THE GAME

Regulation, if it comes, will probably not center on the issue of privacy, but rather on the issue of money.

There are several players in the game. There are, first, the manufacturers of computers (hardware); there are the sellers of data-processing services and programs (software); and there are the communication companies, like AT&T and other telephone companies, who have the network equipment. And the interests of all three are not identical.

The major growth in the new computer industry is in the sale of services over telephone lines. The computer community does not wish to have its growth rate determined by the communication companies.

The evolving issues are beyond the sole domain of the participants. There is the "public" and "the government" to be considered. But the pitiful technical resources of government available to unravel the legal obfuscation of the contestants are small in comparison to the task ahead of them. The budget of that portion of the FCC that must serve the public interest in this new field is incommensurate with the magnitude of this task. As a result, the overburdened staff of the FCC and other regulatory agencies act timidly as the day-to-day protector of *both* the public and the utilities. Allow none to make too high a profit; allow none to jeopardize the profitability of previous investments of the common carriers; disallow the use of income from one set of users to subsidize a new service which may be subject to outside competition, etc. But all this only serves to delay the evolution of computer technology by avoiding those investigations which call for a more complicated regulatory process.

Let us examine some of the economic forces at play in the new computer utility field.

Yearly, the cost of computation equipment declines drastically. A computer system only a few years old is ready for the junk heap. The parts are as good as new, but later models are faster, smaller, and less expensive to operate (per equivalent unit of computation). "Last generation's" computers are too expensive, even though incompletely amortized. Rapid obsolescence imposes a tremendous burden of capital investment upon the computer manufacturer. The more successful it is at leasing computers, the more money it ties up. It is no secret that the computer divisions of most companies are not paying dividends. But the decision of a large electronics company to stay in this high-stake game is not one of choice. The major future element of electronics growth lies in the computer-technology art. To drop out of this game is like the icebox-maker choosing to drop out of the electrical refrigeration business. You either go where technology takes you, or you cash in *all* your chips and leave.

In the unregulated computer business, free enterprise is at its competitive best in stimulating rapid technological growth. The computer consumer is reaping the benefits of large, high-risk capital investment. The tab for rapid equipment obsolescence is picked up to a large extent by the competitive computer manufacturers, not by the consumer.

But let us consider the other partner in the new joint technology of computers and communications. Communications between the unregulated computers and their remote users is over circuits provided by regulated utilities such as the telephone company. The costs and tariffs in the regulated communication business cannot be identical, or even commensurate.

To charge each user what his service actually costs would require more accountants (and lawyers to adjudicate differences) than there are now telephone repairmen—and one out of 70 people in this country already works for a telephone company or its suppliers! Moreover, the potential spread between communication costs and tariffs widens as communication itself is simultaneously undergoing technological improvement—especially in systems designed primarily for the transmission of digital data. Although such systems exist today only as paper studies, it may not be understating the case to say that the full momentum of this new technological force is blunted by the inertia of present government regulatory policy, which isolates tariffs from the type of equipment used to provide a service, and limits the type of equipment that may be connected to the line.

THE IMBALANCE OF COSTS

A period of rapid technological innovation can produce temporary dislocations during which obsolescence can occur before amortization. The end result will be markedly different, depending on whether the industry is regulated or unregulated; and, if regulated, whether the utility is free from competition or from substitute processes.*

In the real world, as costs and tariffs of communication are (almost by definition) not usually identical, only a long-term decline in tariff rates may be expected. But if the change in technology is sufficiently abrupt to allow a new entrant to create an entirely new communications system, or a portion of a system offering a lower price than tariffs based upon the amortized cost of existing equipment, then a new range of possible alternatives presents itself. Some of these could be violent. (There are some who hold that because of development in digital technology, microwave, and satellites, it will be possible to build entirely new special communications systems designed solely for digital communication, at a lower cost than is represented by existing tariff charges.) Present data tariffs are based upon using a communications plant designed primarily for voice communications. Today's limited number of data subscribers are minority users of the system. Digital transmission and switching is being adopted by the telephone company in their future plans as the most economical way to serve new voice and personal television users. But the full advantage in switching to digital communications techniques is not felt by the mass of huge, slowly-amortized telephone plant which is used in the overall tariff calculation. This being so, the plunging cost possible (in a system designed for data transmission or at least optimized for data transmission) does not result in

---

* Computer manufacturers try to write their equipment off as fast as possible. Contrast this with the seemingly strange behavior of telephone companies that refuse to use the newly allowed seven-year writeoff permitted by the Internal Revenue Service —preferring a twenty-year writeoff. The reason is that the legal tariffs are established by the book value of the plant. You can charge the customer more if you are a regulated utility and if you can show a larger unamortized plant investment.

major savings to the computer user. There are some companies eager to enter the data communications field, but holding back because of the legal restriction to competition—and a well-founded fear that the telephone company might recalculate its tariff offerings and wipe out the new competition.

Only a few years ago this would have been of minor importance. The data users were few, and communication cost was low compared to computation. But now the cost for computation is dropping so rapidly that, relative to computation, the cost of communication needed to tie together many users of a single machine is the overriding cost limitation. A few years ago, the cost of communication was only about 10 to 15 percent of the total computer-communication system cost. In some of the larger systems now being considered, communications cost is 50 percent or more of the total system cost. Estimates of over 60 percent are anticipated for some later time-shared systems.

This increasing imbalance causes the computer user to press the communications suppliers to reduce their tariffs. This desired price reduction has, in the past, only appeared as a polite request: it was hardly worth any single computer company's time to press the issue, especially since all computer suppliers paid the same price and no one reaped a competitive advantage. But in the impending evolution of time-sharing systems, certain developments could bring these gentlemen's agreements to an end, and a donneybrook might result.

THE NEW TECHNOLOGY OF DATA CONCENTRATION

It is both technically feasible *and* necessary to concentrate the traffic from a large number of separate users in order to share expensive transmission circuits. Each telephone circuit, plus a little storage to smooth peak loads, is capable of handling as much as a few hundred times the *average* load generated by first-generation remote computer terminals.

Obviously, if communications costs continue to be overriding, there is a tremendous payoff for "concentrating" a large number of users together over a single telephone circuit and splitting the bill among them. Today, as has been observed, a small time-sharing system cannot economically serve customers more than 50 to 100 miles away. But if many independent people in Los Angeles wished to share a single communications circuit which would tie into different computer systems in the Boston area, it is perfectly feasible for them to band together to build a small computer-like device—called a "digital concentrator"—which gathers all the signals together and transmits them over a *single* circuit, leased from the telephone company. *The cost of communications per user would become so low by this arrangement that it would be possible economically to "talk" to a computer almost anywhere in the country. There is only one catch—this may be in violation of present-day laws! We would have created an unregulated common carrier.*

Today we can concentrate only if: (1) we have enough users served by the *same* computer company to divide the costs of the concentrator; (2) the communications are used *only* for computation; and (3) we agree not to exchange "messages" between users. In other words, if all the computer companies could get together, this would increase the economic area that can be served by a time-sharing system from 75 miles to over 3000 miles.

But this would be illegal unless all the users, or all the end computers, were owned by a *single* company. One could imagine IBM rapidly achieving this condition. But the smaller computer companies cannot share communications facilities to compete for distant users. If they did so, they would become illicit common carriers—unless they were regulated. And there is little precedent upon which to create such new regulated common carriers. The legal roadblocks anticipated, if the smaller computer companies sought this avenue of relief, could jeopardize their activities for years.

ONE OR MORE COMMON CARRIERS?

The creation of a new "concentrator" service appears indicated. But there are difficulties with this approach.

Although the telephone company has expressed no interest in doing so,* there are computer manufacturers willing to enter the field *provided* that they are protected from the uncertainty of the telephone company later entering the business, after the market is "proved out," and before their special-purpose equipment is amortized.

The strongest barrier to the proposed common concentration approach is that the telephone company itself may be expected to raise strong objection on the legally correct grounds that one has created a new common carrier which apparently would not be subject to regulation. Whether this objection would vanish if this "new communications utility" were to be regulated in some manner, it is difficult to predict. While the telephone company would temporarily lose business because of such concentration, it would still be the supplier of the basic links required by the new regulated communications service.

Another cloud of doubt is the possible enforcement of the ancient Western Union Telegraph Company franchise as *the* carrier of "record traffic"—that is, of *written* text as against oral communication. Some lawyers hold that, even though the computer did not exist at the time of the granting of the Western Union franchise, if the computers exchange "record traffic," Western Union's legacy from a dead era gives it a case upon which to argue that it alone can legally provide the desired services.

This viewpoint is not willingly accepted by most computer manufacturers, and there is strong opposition to the exclusive control of this potential market by the telegraph company. After all, Western Union is already in the time-sharing computer services market—it is a competitor as well as

* Or some people say "can't," according to their interpretation of the 1956 AT&T-Western Electric antitrust consent decree.

supplier. Further, few computer companies wish to see the rate of technological development of their industry hinge on any *single* organization.

Assuming the maintenance of the present regulatory pattern, we may still expect to find data concentrations being developed. But their growth pattern will be different. The small time-sharing company will not be able to serve enough users to do very much concentration economically, except within a few major cities. Conversely, the large computer companies will serve a large number of customers earlier in the game. As such, they will be able to provide an earlier and broader concentration, and so reduce the communications cost—to their own users only.

As the cost of communication will be expected to be the overriding consideration in many instances, merely by standing still we shall allow a situation of dynamic instability to occur. The larger computer utilities will grow larger as they service the entire nation economically. The smaller companies will suffer. Such ineffectual "regulation" will drive out the smaller "computer utilities." While this would greatly simplify the burden of regulation, we should appreciate the secondary price we may pay and should carefully consider the alternatives.

POSSIBLE POLICY MEASURES

The criteria for the choices facing us are as follows:
1) Aid the development of the new computer technology;
2) Offer maximum protection to the preservation of the rights of privacy of information;
3) Encourage the development of competitive enterprise in all sectors of the new "utility," recognizing that preservation of free choice is not only economically feasible but desirable;
4) Minimize the complexity of the regulatory process so that it does not stand as a potential hindrance to new entrants.

To these ends we might wish to consider the following policy choices:

1) Initiation of Professional Licensing Standards. To deal with the problem of privacy, and to insure that there is no infringement by computer technicians on personal or business secrets, we may have to face the problem of licensing. The professional-technical societies whose members include programmers, design engineers, technicians, and managers of the new computer utilities must first confront the issue. An early inquiry into standards of ethics, fiscal responsibility, and penalties for abusing the proposed licensing standards would provide a basis upon which to construct privacy protection schemes. Although licensing would probably be performed on a state-by-state basis, a model National Code would be helpful in expediting reciprocal licensing agreements. As data are already flowing on an interstate basis, the possibility of national licensing should also be considered.

2) Remove the Economic Advantage to Large Computer Utilities by Changing the Ground Rules on Concentration. Unless the rules prohibiting smaller computer utilities from gathering together their traffic from remote cities and concentrating it for transmission are modified, we can expect the larger computer utilities to drive out the smaller ones merely by virtue of lower communications cost. While this problem can be solved by the creation of new common data carriers, the price and time delay of the inevitable litigation may kill off the smaller companies before their cases are adjudicated. Either simpler licensing procedures or redefinition of "common carriers" is indicated.

3) The Right of Free Exchange Across System Interfaces. A new pronouncement by the regulatory agencies of a doctrine of free interchange of signals across the boundaries of individual systems would be of tremendous technological benefit. Once, when the telephone company sold *only* a mouth-to-ear service, constraints on all equipment connected to the system may have been justified. But, today, the telephone plant is carrying a heterogeneous mixture of signals. The nature of the signals should be of no concern to the telephone company (provided these signals do not interfere with one another). The user should be allowed to connect any mixture of computer and other signals without having to pay a different tariff depending upon who is sending what.

4) The Right of Attaching Foreign Devices. Telephone companies have jealously guarded their right to control all devices attached to the telephone system by a doctrine euphemistically called the "foreign attachments doctrine." They have even hauled into court those who would issue free plastic binders for telephone directories. Today, no computer can be connected to the switched telephone system unless a box called a "Data-phone"* is inserted between the telephone line and the computer. The computer manufacturers feel that they could reduce the cost of their remote terminals if they were allowed to build their own circuitry (and still protect the telephone from a potential source of interference). This would mean fewer pieces of equipment on a customer's premises requiring service by another organization. Computer manufacturers would like to receive a definitive standard of the electrical signal required to feed telephone lines. They feel that a telephone utility should provide communications, and not data-processing, equipment.

5) Encourage the Use of Radio for Data Transmission. The FCC could stimulate competition in the local communications field by encouraging the use of the lightly occupied Super High Frequency spectrum for the transmission of data. In the future, there will be a growing need for a single computer (or concentrator) located in the center of a city to spurt data to inexpensive TV-like displays at a large number of remote terminals. Highly directive microwave offers an economical solution to this requirement—provided the licensing red tape is reduced and we allow some newer forms of signal modulation to serve a large number of highly intermittent

* A registered trademark of A.T.&T.

computer users who share a common frequency band. FCC rules and standards here delay the development of new communications technology.

We are moving headlong into an era when information processing will be available in the same way one now buys electricity. But the attributes that distinguish heterogeneous information from the homogeneous products distributed by the earlier utilities demand a fresh regulatory approach—one that would serve technological progress, privacy, and the public interest. We may wish to think of this as tomorrow's problem; but it is already here.

The Credit Card Society

ALLEN H. ANDERSON

University of British Columbia

The emergence of a checkless society appears imminent, with only a few developments necessary before such a system will be operational. Most of these developments will require only a refining of present knowledge. No major technological or economical breakthroughs are foreseen which will have to precede the successful evolution of such a system.

In brief, four basic technical requirements will be needed to execute electronic funds transfers in the checkless society. First, small, inexpensive, and reliable terminal devices must be located at the point of purchase or site of transaction. Second, massive random-access computer files (containing information, for example, on individuals' current deposit balances, credit information, or both) must be operational. Third, there must be a fast and efficient communications network linking terminal devices with central data files. The use of a central switching computer for linking systems participants will be necessary to facilitate the proper flow of information throughout these networks. Fourth, . . . a machine-readable, unique, identification card for each individual will be required to activate the transaction system. . . .

It seems appropriate that the checkless society be placed in perspective. It should already be evident to the reader that the described system signifies the greatest innovation in the history and evolution of money in this century. Reflection on the past, however, shows that other major innovations, with similarly far-reaching consequences, have occurred on numerous occasions.

The original barter system, for example, evolved several thousand years ago when primitive man realized that to satisfy his personal need for goods and commodities, trade with his neighbor was worthwhile and necessary.

Reprinted from Allen H. Anderson, et. al., *An Electronic Cash And Credit System* (New York: American Management Association, 1966), pp. 16–18, 119, 122–132, by permission of the American Management Association.

However, as man further developed and his societies became more sophisticated, his economic needs became more diverse and complex. The barter system was often cumbersome. The trader really needed some sort of common medium of exchange, something that was easy to carry, scarce, divisible into units, acceptable to all tastes within society, and, finally, applicable for use in different types of trade. This need was soon satisfied by the innovation of a medium of exchange that was recognized by all as being symbolic of value. Gold and other rare metals that could be stamped into coins of standard weight soon began to serve these functions. As international trade expanded, even the fair exchange of rare metals and coins became an acute problem. International merchants must have spent endless hours figuring the value of their various stores. In the more complex and specialized society that finally evolved, the need for a better value exchange mechanism was satisfied by a new innovation—banks. Alert middlemen came to expand their money services to include real banking operations whereby clients of these "bankers" simply deposited their coin and gold for safekeeping. In return, written receipts were issued to the clients. Since these receipts symbolized value and since society needed a medium of exchange that was less awkward than heavy coin, another innovation in the evolution of money occurred—the use of paper as a medium of exchange. Then, of course, came the check and the banking system we know today.

Now, with over 60 million checks being written each day in America, the need for reducing the growing time and expense in handling this paper is a stark reality. It is the authors' contention that by 1985 (when the number of checks processed daily would otherwise have exceeded 100 million) the check will have become as obsolete as the barter system and that a universal "funds identification card" will have become the new, if not the ultimate medium of value exchange. . . .

In the evolution from today's system of funds transfers to the [electronic cash and credit] system, . . .many changes in the legal aspects of banking, credit, and retail operations will occur. . . .

There are two [major] legal complications which could result from an electronic system . . .—antitrust implications and fair-trade problems.

The source of control for an integrated cash and credit system with its required equipment is discussed later on, but it will certainly be so expensive to install, to place into operation, and to operate that a parallel, duplicate system, or systems, most probably would not be feasible. This rules out competition and logically leads us to consideration of an authorized monopoly or integrated "computer utility." Government intervention in the form of rate setting, control of assets utilization, and so on, would, of course, be the logical next step.

Further, there will almost certainly be an effort made by the banks, and possibly by the retail institutions, to control the credit function. The possibility of this occurring will be related directly to the speed and force with which the credit bureaus apply themselves to establishing a place within the system and to the attitude of the antitrust division of the Attorney General's office toward such an action. No prognosis can be given except in the instance where the antitrust division refrains from acting and where the credit bureaus do nothing or too little. This instance would surely result in the credit bureaus' demise.

In the area of fair-trade practices, Government intervention will surely result unless utmost care is taken throughout the evolution to insure equity in establishing and applying the rate structure for data transmission, in accessing the computer, and in permitting purchases of the additional services made available by the system.

CONTROL AND OWNERSHIP OF THE ELECTRONIC CASH AND CREDIT SYSTEM

The question of who will own or control the nation's electronic cash and credit networks looms as one of the most important considerations for systems planners. Robert V. Head, formerly of Touche, Ross, Bailey & Smart, has discussed in a paper, "Emergence of the Checkless Society," the alternative economic sectors which will have a vested interest in the operation of the system. Much of what he said is incorporated in what follows.

Banks. Commercial bankers. . .appear to be in the most logical position for developing cash and credit transfer systems capability; it is within the banking sector that the most advanced applications of OLRT [on-line real-time] systems for funds transfer are now to be found. Banks are clearly demonstrating the resources both for setting up the transfer mechanism and for supporting sufficient consumer loans to extend credit at the point of a retail transaction. Commercial bankers are regarding the electronic funds transfer systems as logical extensions of the other fiduciary services they are performing, and it is to be expected that banks will seek to maintain a strong hand in the control of the new systems.

Credit Bureaus. Credit bureaus, seeking to meet the demands of their users for more rapid, comprehensive, and accurate data, are likely to offer OLRT services of their own with interface to merchants and banks. As an on-line participant in the system it is natural to expect that the credit bureaus will also be extremely interested in who controls and operates the system, to insure the protection of their own information files and of the clients whom they serve.

Retailers. The retail merchants in some areas may be unwilling to stand by idly while the system is introduced by some other economic sector. In fact, it is suggested by Mr. Head that retail merchants might band together in a community to set up credit transfer systems of their own, perhaps pooling accounts receivable as an evolutionary prelude to a full-funds transfer system. That the retailers might precede the bankers seems unlikely, but, in cases where this is so, it must be recognized that the system developers will be very reluctant to yield controlling authority to another group, such as the banks.

Hardware Manufacturers. Equipment manufacturers might well view the emergence of funds transfer systems not only as an opportunity to supply substantial amounts of equipment but also as an opportunity to secure the substantial revenue that will accrue to whatever organization is successful in providing the service. The growth of OLRT service bureaus similar to the Key Data Corporation may well become organizational extensions of the manufacturers. If this becomes the case, it is possible that the manufacturers will emerge as a powerful element in the cash and credit economy of the future.

Communications Companies. The communications common carrier companies, like the hardware manufacturers, might also contemplate a shift in their established role. Firms such as American Telephone and Telegraph and Western Union are already heavily involved with the development of computer-oriented data handling systems of all types. They could logically arrive at the conclusion that they would be willing not only to provide the needed data links and possibly the terminals but also to offer a full-scale service for electronic funds transfer. Even with overall control vested in another sector, it is clear that the communications common carriers will have a very large voice in control of the telecommunications aspects of the system in matters such as data transmission charges, transmission rates, and universal touch-tone network installation rates.

Finance Companies. Some of the major finance companies presently have an advantage even over commercial banks in that many of their systems are already national in scope and are able to offer the universality so important to the funds transfer mechanism. The existence of a well-established national organization, added to the finance companies' capability to provide consumer credit, places these institutions in an excellent position to establish and control their own funds transfer systems. Like the retail merchants, if this sector becomes the initiator, it will not be inclined to yield controlling power to other organizations. Though such a development is possible, it does seem unlikely at this date that the finance companies will be the initiators of a major OLRT network for transferring funds and credit to facilitate retail shopping.

Independents. Independent entrepreneurs with vision and substantial capital backing could well be the first to launch electronic cash and credit systems in communities where the established organizations, in particular the banks, fail to move aggressively to exploit the opportunities inherent in the OLRT transfer networks. The controlling power of any such entrepreneurs would be substantial—equivalent, say, to the power of a local bank in terms of credit-line extension and the like.

Government. The Federal Government will inevitably have a strong controlling influence in the system, regardless of who has primary control at local levels, because of the far-reaching implications of such a system for various Government agencies. However, there is no assurance that the Government's concern will be limited to regulation of the way in which these systems evolve and operate. It is altogether conceivable that the Federal Government may be compelled .to become the owner and operator of these systems as well. The Government's participation in the Federal Reserve System, in check processing, and in the postal system indicates clearly that the Government's present involvement in funds transfer is substantial and that it could be developed considerably further.

From what has been said, it is clear that in the first years, at least, we can expect no uniform pattern of control to emerge. Problems of interface, compatibility, and systems redundancy, to name a few, will resolve overtime into both standardization of components and emergence of the best systems and most capable operators. As already suggested, we believe that banks as a group will emerge as the primary initiator and controller of future systems, with a strong regulatory influence from the Federal Government; an ownership and control structure similar to COMSAT, with joint Government and private-sector participation, may very well be the ultimate outcome, as suggested by many who have examined this problem. The alternatives to allowing the Government to have a substantial amount of participation could allow the business of credit information and funds transfer to fall to a very large national corporation, with all the resulting consequences of monopoly control. The resolution of the control and ownership questions will become increasingly complex as more and more organizations perceive opportunities to become part of the system; this aspect of the checkless society will undoubtedly draw increasing amounts of attention and concern in the years immediately ahead and warrants careful study by those venturing to establish their own position in what is to come.

THE IMPLICATIONS FOR THE FEDERAL GOVERNMENT

Robert Head identifies a list of at least 12 Federal agencies that will need to be intimately concerned with the emergence of the electronic cash and credit system because of the effect the system will have on each of their activities. The two most obvious are the Treasury Department and the Federal Reserve System.

The Treasury. Within the Treasury Department, the Office of Comptroller of the Currency has control over Federal bank regulations. . . . Many accommodations in current bank regulations will be called for as large funds-transfer networks are established. The Internal Revenue Service will be concerned for many reasons, including revenue forecasting and collection as well as taxpayer identification. The electronic system will also require smaller amounts of currency and coin in circulation, directly affecting the planning and operations of the Bureau of Engraving and Printing and of the U.S. Mint.

The Federal Reserve. The Federal Reserve System, responsible for establishing credit regulations, operating interbank data communications, and processing check transfers, will be dramatically affected by the envisioned system. On February 9, 1966, George Mitchell, a member of the board of governors of the Federal Reserve System, delivered a paper on the potential impact of electronic funds transfer systems on the Federal Reserve System and on Federal Reserve float. Mr. Mitchell states that EDP and wire transmission technology has already begun to reduce Federal Reserve float and will, in the foreseeable future, eliminate it. ("Float" is the aggregate dollar amount of checks on any given day for which credit has been passed by Federal Reserve banks and branches to their depositing member banks without receipt of payment from drawee banks, less the amount not yet given.) During 1965 the average daily amount of Federal Reserve float outstanding was $1.8 billion, amounting to a huge extension of free credit to banks and depositors. In the same year the Federal Reserve banks and branches handled over five billion checks and other cash items having an average daily value in excess of $7 billion.

This huge amount of float has been kept in check only through the installation of MICR [magnetic ink character recognition] and EDP [electronic data processing] technologies, which allow checks to be processed through the Federal Reserve System at the rate of some 60,000 per hour versus a previous 1,500 per hour.

Wire transmission technology is now suggesting further reductions in, if not the total elimination of, float through one of two processes, both of which are currently under study by the Federal Reserve System. The first, a near-range program, would call for the immediate crediting via communications networks to the reserve account of the depositing bank of all checks deposited for collection in Federal Reserve banks and branches, with a simultaneous charging to the reserve account or correspondent account of the drawee bank.

The second process involves a long-range prospect and is the system of cash and credit transfer proposed by the authors, which would eliminate the use of the check itself for the bulk of regular money settlements. In the checkless society as explained earlier, the payor initiates the settlement process by communicating, not with the payee in the form of a check, but with his own bank, notifying it directly whom to pay and how much. This concept, when applied to transactions which pass through the Federal Reserve System, eliminates both checks and float.

Mr. Mitchell suggests that the process of instantaneous settlement and deposit accounting for funds transfers between banks and for bank deposits could be carried out by between 250 or so computer centers located throughout the country.

The structural and organizational changes in the Federal Reserve

System to accommodate such a completely new system are staggering. How-ever, the possibility of eliminating the $1.8 billion in daily float provides a major incentive for the Federal Reserve System to develop an electronic cash transfer as soon as possible. Float is also subject to widely fluctuating volume levels from day to day, because of snowstorms, floods, and other conditions. These fluctuations present serious operating problems to the Federal Open Market Committee in its attempt to maintain bank reserve positions at agreed-upon levels. Surely, any technological developments which will assist the elimination of this problem would merit close study by the monetary authorities.

Other Agencies. Some of the other Government agencies suggested in Head's paper which will be affected by an electronic cash and credit system (perhaps to a lesser degree than the Treasury and the Federal Reserve Sys-tem) include the following:

The Bureau of the Budget, with its national budgetary planning respon-sibilities, will recognize considerable shifts in the levels of consumer credit and other affected variables which may require modification of economic reporting and planning procedures.

The Federal Communications Commission will be heavily involved in the data transmission requirements of the proposed system, both as a regu-latory agent and as policy-forming body.

The Council of Economic Advisers will surely be interested and con-cerned with the overall economic implications of the new system.

The Department of Labor will have to monitor the employment impli-cations of job displacement and new-job creation and may even become involved in manpower planning for the system.

The Department of Health, Education and Welfare should be interested in the system's effect on the individual, not to mention the possibility of making automatic welfare payments through the system.

The Federal Deposit Insurance Corporation will be concerned with the effect the new financial system will have on individual bank balances and levels of bank credit.

The Department of Justice, Antitrust Division, will naturally be caught up with the antitrust considerations. The Federal Bureau of Investigation will likely be interested in the means of customer identification and inter-state law enforcement.

The Department of State might well be interested in the implications of a worldwide complex of automatic funds transfer systems.

The Post Office Department will be interested in the impact of decreas-ing amounts of business and consumer mail and the resultant effect on postal revenues. The complete obsolescence of the postal money order might be implied.

Thus the effect of electronic cash and credit systems upon our national economy has to be evaluated carefully by many agencies. It is clear that the checkless society will affect many more institutions than just the com-mercial banks.

SOCIAL IMPLICATIONS

Of all the complex adjustments which will have to be made before an electronic cash and credit system can become fully operational, some of the most difficult changes must occur in the areas of human behavior, attitudes, and work patterns. The common element of all the problems to be discussed is the problem of change itself: Unless there is a distinctive "gain" involved in the replacement of a habitual pattern of activity with a new one, individuals and groups are naturally reluctant to make the change. When other specific sources of resistance, such as regional parochialism, fear of direct competition from new or different types of economic units, or fear of anything new, are combined with the general resistance to change, the process of winning acceptance for the proposed transfer system may be very difficult indeed. The major areas where these problems will occur are as follows:
1. The manning and operation of the system.
2. The reactions of individuals to the new method of making transactions.
3. The resistance of several groups of businesses on economic grounds.
4. The impending threat of "big brotherism."

Manning the System. Finding, training, and maintaining qualified people to operate the system will be no easy task. There are already serious shortages of critical-skill workers in the occupations related to computer activities, such as systems analysts, programmers, even key-punch operators. Although the hardware manufacturers, the larger firms using computers, and the privately operated schools are attempting to train large numbers of people for these jobs, supply is still running far behind demand.

A significant increase in demand for computer technicians can be expected when banks, credit bureaus, and retailers plan electronic cash and credit systems, not simply because the programming involved is complex but because identical programming tasks will probably be duplicated many times in sort of hodgepodge development of regional centers. Even if the hardware manufacturers develop general-purpose programs to control cash and credit transfers (as they have for other applications), each center will still have to adapt the program to its own unique application requirements demanding the time of technicians skilled in computer operations.

Within each regional center, the programming required to tie each unit into the central switching computer will require thousands of man-hours of work. Competing businesses within a region, such as two banks tied to the same central switcher, are not likely to share programs; even today bankers frequently view their "own" programs as superior to functionally identical programs used by competitors. It's all part of the competitive spirit, perhaps. Competitive aspects of computer utilization (and the corresponding cost of redundant programming and systems work) will likely become more pronounced before it gets better as units within the system each hasten to be the first to develop and offer new types of computer-based services to utilize their OLRT capability.

The shortage of technicians, which may continue far into the future, will require more than the basic economic mechanism of supply and demand to provide a solution. The problems of educating young people as to the importance of the computer's existence, of impressing upon older people the need for continually upgrading their skills, and of encouraging the un-skilled to seek opportunities in new industries or areas will all become so urgent that increased Government activity can be expected to help change basic attitudes and provide training.

A more subtle aspect of the same problem, which cannot be solved by Government-sponsored training programs or Government-granted economic incentives, is the difficult task of updating management's skills in computer usage. It will not be possible for the electronic cash and credit system to get off the ground until those managers responsible for making the required investments are also capable of using the system effectively. This implies that managers will need to develop both increasingly higher levels of sophistication regarding computers and also more detailed technical knowl-edge in terms of managing the computing operation.

Reactions to the New System. Three problems relating to individuals' present transaction habits may impede customer acceptance of electronic funds transfers. With the checking system, customers have the ability to stop payment on a check if they are dissatisfied with the merchandise they purchased. This customer expectation cannot easily be reconciled by the proposed system. What could be done technically, however, would be to include a system feature permitting a type of payment delay. Purchasers of relatively expensive items could delay actual payment for a specified period (at which time the demand deposit transfer would take place automatically) in exchange for payment of a flat or percentage-based "interest" fee. Since the number of stop-payments is not significant in relation to the number of *transactions,* the social resistance to loss of the privilege could easily be overcome by an educational campaign to inform customers of such optional methods of withholding or delaying payment.

A second problem which has attracted the attention of pessimistic students of American social behavior is that of the possibility of a "credit spree" occasioned by the sudden ease with which individuals could spend up to the limits of their lines of credit under the proposed system. While the recent dramatic increases in consumer credit have been viewed by some as an indication of a lack of financial responsibility (even the degeneration of moral values), it does not appear that the increased convenience of using an identification card versus writing a check or arranging for credit at a store is so great as to cause a revolutionary shift in credit-purchasing habits.

Another potential problem will be reactions to the loss of check float which all present users of checks will experience. The time lag between the writing of a check and the actual deduction of that amount from one's checking account balance has been used to advantage by groups ranging from students, whose checking account balances may ride perilously close to zero, to huge corporations, which use the time value of money to make more money. While the effect of the elimination of float may be mitigated in the case of businesses, whose present check transactions take place both ways, those individuals who could not maintain positive account balances if all outstanding checks were suddenly posted, will be forced to revise their purchasing and payment habits or pay the interest charges on short-term extensions of overdraft credit. While this may be of concern to experienced practitioners of the art (kiters most notably), it is felt that the general public, including those who are not aware of the time involved in processing a check, will be able to forgo the benefits of float in exchange for the extra convenience of instantaneous transactions and possibly the elimination or reduction of charges for checks and account servicing. Many people never even consider the concept of float anyway and *assume* that checks they write are instantaneously deducted from their account.

Business Resistance. Managers of smaller banks and retail stores have voiced the fear that the installation of an electronic cash and credit system will cost them their identity—that they will be "swallowed up" by the larger units within the system. Smaller banks which cannot afford large computer installations have worried that the control of the electronic system might rest entirely with the large banks; also, since frequent personal contact with depositors will decrease once deposits and withdrawals can be made from remote locations, the controlling banks would be able to out-market smaller banks by offering exclusive services. Actually, the large banks will need to enlist the joint participation of the small banks in order to eliminate the need for processing the checks written by small-bank customers; if only part of the banks and stores in a market area convert to an electronic system, the problem of maintaining duplicate check-processing and electronic-processing facilities mitigates the electronic system's advantages. One of the beauties of OLRT technology is that small banks will be able to market all the computer-based services that the large banks will offer, by sharing computer time with the larger banks or by joint ventures of smaller banks, *without* a disproportionate investment in computers and equipment.

. . .Many retailers fear the elimination of their own credit-granting facilities will be a loss of an important marketing tool. They claim their present credit cards provide an exclusivity that is worth more than the cost of operating their credit-granting services, that their credit standards are more liberal than the banks' and therefore attract customers who would not receive credit elsewhere, and that the loss of credit-granting capability would actually cost some of the large efficient retailers the profit they make on credit operations. A more general concern with retailers is that they have not yet had sufficient years of computer experience to appreciate fully all advantages of the OLRT system. Many retailers are still busy resisting the basic concept of computerization and using computers to help in their internal operations. . . .

Big Brother. The last, most serious social problem which the new system raises is the question of control not only of the system but of people. There are viable, convincing arguments to be made for integrating the functions of banking, retailing, and credit through electronic networks. One of the natural results (and advantages) of doing this would be to consolidate the information about individuals and thereby broaden information "coverage." But unless the development of proper controls over this potentially huge source of personal information precedes or accompanies the building of the system itself, confusion over the authority to use information will undoubtedly result; and possibilities for unscrupulous use of data might arise.

Even if we assume that the security safeguards in existence now can be kept intact and that improper commercial use of information will be impossible, there still remains the question of possible governmental use of the collected personal data. It is conceivable, for example, that deterministic applications of governmental standards could affect opportunities for individuals to qualify for government or civilian jobs, hold elective or appointive office, or enter advanced publicly owned schools. It is possible, of course, to conjure up an Orwellian world of complete governmental surveillance and control, all made possible by what started out to be an efficient way of transacting business!

The authors do not believe, however, that we need have any deterring fears of 1984 when we consider the automated system proposed. In the first place, the individual's right to privacy has historically been of such importance to the makers and interpreters of law that the present restrictions upon usage of personal information will never be relaxed, nor will the system of checks and balances which controls the relative power of interested governmental groups ever change to the extent that one group or agency could dominate control of the system. The other, more basic reason, is that the information that would be available within the proposed system is exactly the same information that is being collected now. All this information *could* be gathered (most of it legally) by anyone or any agency *today* that wanted to do so.

The internal reorganization of banks' files to build central information files, the development of computerized credit files, and the establishment of customer files in retail organizations have not triggered a rash of speculation that "big brother" is upon us. The authors can forsee, therefore, no reason for believing that the simple connection of these files through on-line data links will do so either.

The Leisure Society

ERWIN O. SMIGEL

New York University

One of the major assumptions concerning automation has been that it will bring with it drastic reductions in the number of hours we need to work. A recent survey of the literature on automation,[1] however, indicates that the drastic reduction predicted, since John Diebold coined the word "automation," has not yet occurred. Still, no one feels that a major reduction in working hours will not come about. In *America's Needs and Resources,* J. Frederick Dewhurst estimates that by the year 2,000 the average workweek will be fifteen hours.[2] If this decline does take place, leisure time will, indeed, present a problem. The impact of leisure on American industrial society and upon industry itself is becoming one of our new concerns.

However, before we discuss the ramifications of the increase of leisure, we must know a little more about it. Hard facts about leisure are not easy to come by, and information concerning the interrelationship between work and leisure is even more difficult to find.

The scarcity of both definition and interpretation is perhaps in large measure attributable to the Calvinistic principle prevalent in American culture—work alone is good; a preoccupation with leisure borders on an endorsement of sin. To illustrate, in 1926 John E. Edgerton, president of

Reprinted from Erwin O. Smigel, "The Problem of Leisure Time in an Industrial Society," Industrial Relations Monograph No. 25, *Computer Technology—Concepts For Management* (New York: Industrial Relations Counselors, Inc., 1965), pp. 103–120, by permission of the author and Industrial Relations Counselors, Inc.

[1] Anne Fonner, "Automation: Its Meaning for Sociology—A Survey and Analysis of the Literature," unpublished Master's Thesis, New York University, New York, 1963.
[2] J. Frederick Dewhurst, *America's Needs and Resources, A New Survey,* New York, The Twentieth Century Fund, 1955.
Note: Portions of this paper were previously published in the introduction to Erwin O. Smigel (ed.), *Work and Leisure: A Contemporary Social Problem,* New Haven, College and University Press, 1963.

the National Association of Manufacturers, said in opposition to Henry Ford's five-day week:

> I regard the five-day week as an unworthy ideal. . . . More work and better work is a more inspiring and worthier motto than less work and more pay. . . . It is better not to trifle or tamper with God's laws.[3]

This attitude has been undergoing a change; in part, the shift reflects growing public concern about the work-leisure complex. The concern stems from a fear of unemployment that had its roots in the 1929–39 depression, with its enforced "leisure." Automation, and its threat and promise of increased free time, has intensified the general awareness of the subject, and the unions' demands for a 35-hour week have kept the discourse current.

Sociological theories dealing with the dysfunctional use of free time are beginning to be discussed outside academic walls, and one occasionally hears public comments on the proper and improper use of leisure time.

WHAT IS LEISURE?

The study of leisure itself involves some unsolved problems. The first difficulty originates from confusion over the meaning of the word. Often the term "leisure" is employed when it is the "use" of leisure that is actually being discussed. And the confusion is compounded when words such as "recreation," "amusement," "fun," "pleasure," "play," "hobby," "idleness," "free time," and "leisure" are used interchangeably.

Two major schools of thought have emerged concerning the definition of leisure. The first is headed by political scientist and philosopher Sebastian de Grazia, who champions the definition of the ancient Greeks. He offers as the key to the meaning of leisure: "Freedom from the necessity of being occupied," which paraphrases Aristotle.[4]

A poem published in *The Reporter*[5] is a succinct statement of de Grazia's philosophic position on the proper use of leisure.

> Lie down and listen to the crab grass grow,
> The faucet leak, and learn to leave them so.
> Feel how the breezes play about your hair
> And sunlight settles on your breathing skin.
> What else can matter but the drifting glance
> On dragonfly or sudden shadow there
> Of swarms aloft and the wiffle of their wings
> On air to other ponds? Nothing but this:
> To see, to wonder, to receive, to feel
> What lies in the circle of your singleness.

[3] Russell R. Dynes, et al., *Social Problems: Dissensus and Deviation in an Industrial Society*, New York, Oxford University Press, 1964, p. 285.
[4] Sebastian de Grazia, *Of Time, Work and Leisure*, New York, The Twentieth Century Fund, 1962, pp. 12–15.
[5] Smigel, *op. cit.*, p. 2.

Think idly of a woman or a verse
Or bees or vapor trails or why the birds
Are still at noon. Yourself, be still—
There is no living when you're nagging time
And stunting every second with your will.
You work for this: to be the sovereign
Of what you slave to have—not
Slave.

The members of the second school mean, by leisure, *free time*. This is the meaning assigned to the term by most modern sociologists. Nevertheless, there is some disagreement about what free time is. Is it merely time independent of work? (If so, how are we to consider forced free time brought about by unemployment or unwanted retirement?) Or, must free time, in order to be considered leisure, be unobligated time, or paid free time? Peter Henle, of the U.S. Department of Labor, thinks that if free time is to have meaning in terms of leisure for workers, it must be paid free time.[6]

There has been a great increase in paid vacation time and paid holidays. Henle reports that at the end of the 20-year period, 1940–1960, the average worker had gained 155 hours in leisure time. Over 50 percent of this increase was the result of fringe benefits—48 hours represented six additional paid vacation days, while 32 hours were the result of four more paid holidays. Henle notes that the one and a half hour decline in the workweek accounted for less than half of the increase.

DISTRIBUTION OF LEISURE

Even if agreement had been reached on a definition of leisure, actual study of the phenomenon remains complex. As Wilensky has pointed out,[7] the availability of leisure differs with the times (and this not always in the expected direction). He reminds us, for example, that over the centuries time spent at work increased before it decreased, and that what is happening now has happened before. Initially, we started with very little time spent at work; work-time was extended when the expertise and equipment became available. Now in a stage of advanced development it is lessening again—but not for everyone.

If we look back at the old Roman calendar of 355 days we find that almost one-third (109 days) of it was marked as unlawful for judicial and fiscal business; the number of Roman holidays reached a peak in the middle of the fourth century when days off numbered 175. And Wilensky tells us, at the beginning of the nineteenth century in France, intellectuals worked 2,500 hours per year, while in 1950 they were working 3,000 to 3,500 per year.

6 Peter Henle, "Recent Growth of Paid Leisure for U.S. Workers," in Smigel, *op. cit.,* p. 183.
7 Harold L. Wilensky, "Uneven Distribution of Leisure: The Impact of Economic Growth on 'Free Time'," *Social Problems,* Summer, 1964. pp. 33–35.

The availability of leisure differs not only with the times but also according to occupation. In the United States, for instance, recent increases in leisure have been unequally distributed, and most of the real gain has come about in private, nonagricultural industries—most markedly in manufacturing and mining. In fact, the old elite leisure class, according to Wilensky,[8] is disappearing and a new leisure class is taking its place. He found that 38 percent of lawyers unaffiliated with legal firms work 55 hours a week, or more;[9] in the managerial class, 33 percent work these hours; among semi-professional people, the figure is only 14 percent, and for other nonmanual workers, 12 percent. Russell R. Dynes and his co-authors[10] claim that since 1962 the new leisure class is made up of clerical workers, craftsmen, foremen and the like, operators of factory machines and vehicles, service workers (except domestics), and laborers, excluding farm. Some 39,307,000 workers are involved in these categories.

LEISURE NORMS

Along with inadequate definition, the work-leisure problem is further complicated by the fact that the ways in which it is used may differ by social class, by nationality and religion, by occupation and by life cycle. And how it is used, affects how it is defined. What once might have been considered leisure need not be considered so today. School, for example, was leisure to the ancient Greeks. Today, for most in the United States, it is not. Compulsory mass education with its utilitarian emphasis has moved learning into the area of assigned responsibility, and thus it becomes work.

One does not, however, have to go back to the ancient Greeks to show that differences have occurred in the use of leisure time. A bulletin posted by a merchant for his employees in 1822 provides a picture not only of "recommended leisure pursuits," but also of work requirements and amounts of free time. The following was displayed in a store in Chicago—Carson Pirie & Scott—and is a representation of the mores of that day.

RULES FOR CLERKS

1. This store must be opened at sunrise. No mistakes. Open at 6 o'clock a.m., summer and winter. Close about 8:30 or 9 p.m. the year round.
2. Store must be swept, dusted—doors and windows opened—lamps filled, trimmed, and chimneys cleaned—counters, base shelves and showcases dusted —pens made—a pail of water and also the coal must be brought in before

[8] *Ibid.,* pp. 40–41.
[9] I recently completed a study of Wall Street lawyers and their large law firms. And my findings confirm Wilensky's, except that I discovered that "my" firm lawyers worked as hard or harder than "his" solo lawyers. Erwin O. Smigel, *The Wall Street Lawyer: Professional Organization Man?* New York, Free Press, 1964, 369 pp.
[10] Dynes, *et al., op. cit.,* pp. 287–288.

breakfast, if there is time to do it, and attend to all the customers who call.
3. The store is not to be opened on the Sabbath Day unless absolutely necessary, and then only for a few minutes.
4. Should the store be opened on Sunday, the clerks must go in alone and get tobacco for customers in need.
5. The clerk who is in the habit of smoking Spanish cigars—being shaved at the barber's—going to dancing parties and other places of amusement and being out late at night—will assuredly give his employer reason to be ever suspicious of his integrity and honesty.
6. Clerks are allowed to smoke in the store provided they do not wait on women with a 'stogie' in the mouth.
7. Each clerk must pay not less than $5 per year to the church and must attend Sunday School regularly.
8. Men clerks are given one evening a week off for courting and two if they go to prayer meeting.
9. After the 14 hours in the store the leisure hours should be spent mostly in reading.[11]

This was the advice in 1822.

While the strict rules of the Protestant Ethic have not all disappeared, new leisure norms are emerging which conflict with the older, more firmly entrenched ones. According to the developing norms, leisure is more than a restorative for work, as it was for the ancient Greeks. Leisure, today, is coming to be considered an important part of life which should be regarded as worthwhile in itself. The problem of dealing with leisure time, however, is becoming increasingly difficult because it is necessary to think differently about it for different work groups, since each has a different amount of it, sees it differently, responds to it differently and, therefore, has somewhat different problems concerning its meaning and use.

DIFFERENTIATION OF WORK INTERESTS AND LEISURE PATTERNS

The problem of leisure, as Bennett Berger indicates,[12] can only be understood by analyzing the work and leisure attitudes of the groups concerned. A number of items are involved in this analysis. One is commitment to work; another concerns the location of workers' primary social relationships. Based on a 1952–53 study in which he interviewed production workers in the midwest, Robert Dubin[13] wrote a prize-winning article in which he tried to determine the extent to which workers were interested in their

11 Delbert C. Miller and William H. Form, *Industrial Sociology*, New York, Henry Brothers, 1951, p. 561.
12 Bennett M. Berger, "The Sociology of Leisure: Some Suggestions," in Smigel, *op. cit.*, p. 23.
13 Robert Dubin, "Industrial Workers' World: A Study of the Central Life Interests of Workers," in Smigel, *op. cit.*, pp. 53–71.

work and in the social relationships they formed on the job. I think that what he found to be true then probably is even more descriptive of the situation today.

The workers interviewed by Dubin were located in different communities ranging in size from 35,000 to 125,000. The interviewees were asked to respond to such statements as: "I would most hesitate missing a day's work, missing a meeting of an organization I belong to, missing almost anything I usually do." Dubin found that only 24 percent of these workers regarded work as their central life interest, and only about 10 percent of the respondents considered that their primary social relationships were formed at work, rather than through social contacts outside of the factory.

Dubin concluded from his study that two influences, contradictory to one another, are operating in today's society. One stems from the fact that work is no longer a central life interest for workers; their life interests have moved out into the community. He notes that much management activity in personnel and industrial relations is directed implicitly toward restoring work to the status of a central life interest. Management's efforts and the main drift of social developments are working at contrary purposes. And Dubin feels that management will not succeed in its endeavor.

The second contradictory influence is caused by the location of primary human relationships in the social fabric. Some groups in management have accepted a philosophy and developed social engineering practices which could be characterized by the phrase "human relations in industry." Again, Dubin feels, the scope must be enlarged; it should be human relations in the community. Although he does not think much of group dynamics, he conceded that establishing such off-work activities as bowling teams, bird-watching clubs, and the like, may be useful.

Louis Orzack,[14] using Dubin's questionnaire, examined the central life interest of professional nurses. He found that 79 percent of them (55 percent more than industrial workers) felt that work was a central interest. This finding may prove interesting in the future if the effects of automation should force jobs to become increasingly professional instead of making them routine. And if these new professionals who work with automation then take on the attitudes of the old professionals, work as a central life interest may again rise to prominence, and the problem for management will be different.

William Faunce studied the work and leisure patterns of automobile workers;[15] he asked them what they would do if they had two extra leisure hours a day. Their responses are important when compared with the ways in which professional people would use their free time. Of the automobile workers, 98.6 percent said they would work around the house, 76.8 percent

[14] Louis H. Orzack, "Work as a 'Central Life Interest' of Professionals," in Smigel, *op. cit.*, p. 75.
[15] William A. Faunce, "The Automobile Industry: A Case Study in Automation," in H. B. Jacobson and J. S. Roucek, eds., *Automation and Society*, New York, Philosophical Library, 1959, pp. 44–53.

said they would spend more time with their families, 53.6 percent said they would travel, 48.8 percent thought they would go to ball games, fights, hockey games, and so forth; 42 percent would go fishing; 12 percent would engage in political action work; 11.2 percent would rest, relax, loaf, and so forth; 4.8 percent would go swimming and boating; 2.4 percent would work on the car; and 1.6 percent would engage in church activity.[16]

Compare these findings with those of Gerstle[17] for professional people and you see major differences. Gerstle asked advertising men, dentists, and professors in Detroit what they would do with more leisure time. Two things are significant: what professional people said they would do differs from what workers said they would do; the responses of the various professional people varied. Of the advertising men, 12.1 percent said that they would relax; 29.4 percent of the dentists said they would relax; and of the professors, zero percent—none of them wanted to relax.

With respect to family and home activities, 24.3 percent of the advertising men, who probably do not spend much time at home, said they would spend extra time with their families, as against 14.7 percent of the dentists, and 9.4 percent of the professors.

With respect to recreational reading, 12.1 percent of the advertising men said that they would spend time in this manner, as against 14.2 percent of the dentists, and 28.1 percent of the professors. The professors are like the busmen, they are going to do what they usually do—read. If you totaled the time the professor spends in reading, or says he would spend, you would find that the sum of recreational and work-connected reading involves 78.1 percent of his time. Gerstle's report gives some indication that professional people, because of their educational background, probably have greater choice about what to do with their leisure time. Workers seem not to have developed outside interests (which they can easily continue in retirement) to the same extent professional people have. Somewhere along the line, this observation may be important to us.

Problems of research in this field are magnified because leisure, being intangible, is often difficult to measure. If a man commutes to work, is the time he takes for travel to be considered work or free time? It may depend on the individual. Does he read the paper, take a drink, chat with friends? It is hard to measure subjective evaluations. To the extent that joy, happiness, and pleasure are involved in leisure, true assessment is difficult.

To sum up, there is difficulty in defining leisure because the definition is not constant. This is so because the word has been used in different ways and has been seen from different vantage points by different groups at different times. Also, since the norms concerning leisure are in a state of change, and the subjective aspects are hard to measure, leisure is difficult to study scientifically.

16 Since many of the respondents gave more than one response, the answers add up to more than 100 percent.
17 Joel E. Gerstle, "Leisure, Taste and Occupational Milieu," *Social Problems*, Summer, 1961, p. 58.

THE PROBLEMS OF ADDITIONAL LEISURE

The problem of leisure has hardly been formulated. It is difficult to imagine what really would happen if we did have a 15-hour week. Of course, some of the problems are easy to see and are already public concerns.

The problems of minority groups which do not have access to play facilities.
The problems of the aged, who are alone and unattended, and whose position in this modern world has suffered a loss of status.
The problems of the unemployed and the handicapped, which are now being studied.

I dedicated a book I edited to my parents, who have just retired. The dedication read, "To my parents, whose work is their leisure." They do not know how to play, and retirement is very bad for them. Their work, in fact, was their leisure, and one of the problems most people will have to face if enforced retirement programs are increasingly adopted is what will happen when they find themselves, at an early age, retired, and not trained for retirement.

Today the general fear of leisure is that we will have time on our hands, and we will not know what to do with it. Professor Robert McIver put it succinctly: "Leisure is ours, but not the skill to use it."

Many intellectuals are worried that some people will waste their new-found leisure in pursuits which have no merit—comic-book reading or, to avoid boredom, association with lunatic fringe groups. This may be a consideration, because if you look at the incidence of juvenile delinquency, you will find it increases in the summer, when delinquents have more time for such behavior or, simply, more time on their hands.

If we check suicide rates we see that they go up on Sundays, when people have more free time.

The problem of what to do with additional free time becomes increasingly significant because the people least trained to handle it will be getting most of it. In fact, just a fast rate of change from work to leisure may lead to discontent and frustration. It is known to do this to those who retire. It logically might occur to those who find themselves working fewer hours. Some workers, not knowing what to do, will take a second job. Moonlighting is, even now, not an uncommon practice. (We even find it among professionals, who work long hours.) For some workers, however, the second job, if it differs from the first, may well be considered recreational. In any event, moonlighting does nothing to relieve what may well be a growing problem of unemployment. An estimated 200,000 jobs will be lost through automation every year for the next ten years, and the loss starts to add up.

Those who do not take the second job often find themselves engaging in exhausting leisure activities—golf at 6:30 a.m., running to the beach and fighting the crowd back, and so on. These are the people who do not know what to do with their time, and who say that they cannot get interested in anything. Many of those who accept the additional free time will find themselves, if they are men, doing housework. Some have already begun to call these extra days off "Honey-do-this" days.

The real danger to our society is that, not knowing what do with ourselves, we will destroy ourselves. David Riesman[18] feels that if we lose creativity in work—as we seem to be doing—we will not have creativity in the use of leisure.

THE USE OF LEISURE

How we use leisure, of course, depends not only upon our training but also upon how our leisure time is distributed. A number of possibilities are being discussed. The day can be shortened so that we have a five-hour day or a four-hour day, and so forth. It might be simpler for society, as it is now constituted, to handle these extra hours than to handle a four-day week.

The four-day week is a second possibility. Rolf Meyersohn[19] examined what people did with the four-day week; he studied employees of a small aircraft parts manufacturing company in California that altered its work schedule so that one week in every month employees got a three-day weekend. One-third of the respondents who initially said they liked the new schedule grew to dislike it. They found they did very little with their free Mondays. Their children and wives were on a five-day schedule and they felt unwanted, in a way. They said they did not enjoy having Monday off because there was nothing to do. To a large extent, free Mondays were used for home chores. If their wives worked, Monday was a lonely occasion without even the relief of television, since on weekdays it is designed mostly for women and children. This, of course, could change if all of us were on the same schedule.

A third possibility is the longer vacation. In fact, the Steelworkers conceived of 13-week vacations, which they call sabbaticals. The union wanted sabbaticals because it saw them as an answer to long-term unemployment. As yet, however, this innovation is not creating any sizable number of new jobs.

What steelworkers will do with this additional time depends on how much money they have and whether they can develop a value system which will permit them to use this long vacation profitably. Only one-half of the men interviewed said they had plans for using the time off. Some union men resent the agreement, because it cuts down on vacation flexibility. The provision requires that all of the sabbaticals be taken at one time.

I think the professor usually has plans to use his vacation. I do not know if this is really so, but I think that professors, knowing when they

18 David Riesman, *Abundance For What?* New York, Doubleday and Company, Inc., 1964, pp. 162–183.
19 Rolf Meyersohn, "Changing Work and Leisure Routines" in Smigel, *op. cit.,* pp. 97–106.

start teaching that they will have long vacations, probably plan ahead for them. When I taught at Indiana University, most professors and their wives thought they would travel when they had their sabbatical. This was the norm. Steelworkers probably have not yet developed norms to guide them in the use of their long vacations. And under some agreements where workers have a choice between a 13-week vacation and a payment into a company savings plan, a majority are choosing the money.

There is a fourth possibility; this is early forced retirement. In each case the problem is similar: what to do with the additional time. With early forced retirement the problem is greater, however, because there is little chance of going back to work and breaking the monotony of unprepared-for leisure. Some companies have plans for phased retirement. In this way, it is thought, a man becomes accustomed to additional leisure time and develops ways of handling it on his own.

We have a very interesting problem now emerging. A number of relatively young men who are officers in the Armed Services are retiring in their early forties. Many are going to Florida—St. Petersburg is one of their havens—and most of these men do not have skills to get other status jobs, or, if they have skills, there are too many of them with the same skills. They do not know what to do with their free time, even though they are on a pension and have money to spend. The Defense Department is a little worried about what is going to happen to these retired officers. And while I do not think industry plans to retire people in their forties, the basic problem is the same. It might be worthwhile to look at the results of the studies that have been completed on these officers.[20]

Each of these four plans to reduce time spent at work presents problems for both management and labor. If historical attitudes continue to exist, labor will remain skeptical of management attempts to do anything about leisure time. They will fear that management will try to undermine the loyalty of the workers to the union. Probably, the best solution would be cooperative ventures between the corporations and the labor unions. However, even in an atmosphere which permitted cooperation, the problems themselves would continue to develop.

Long vacations and the four-day week mean that if manpower is really needed, management will have to deal with more men. The implications of this are obvious: increased bookkeeping, increased training demands, new personalities to deal with, potential new personnel problems, layoff and recall problems and, of course, contact negotiation problems concerned with such details as the form a shorter workweek should take.

If men are retired early, corporations may face the charge that they are not using experience wisely and are wasting training. Even universities face this charge. At some schools professors at sixty-five are forced to retire. Some of these men are both capable and bright, but their wisdom is lost.

If we have a three- or four-hour workday, it may be necessary to increase

---

[20] Albert D. Bitterman, "The Prospective Impact of Large Scale Military Retirement," *Social Problems*, Summer, 1959, pp. 84–90.

shift work. This might turn out to be a necessity anyway, because many people feel that the new machines, if they are to be profitable, have to be kept in constant use.

Edward Blakelock,[21] in a study of the "rotating shift worker," discussed the problems of people whose leisure opportunities are affected by their schedule of work. There was a stress on the restriction of social opportunities. Workers felt it would be difficult for them to participate in activities if they could not count on being free on a regular basis at the same time as others.

Blakelock found that the shift workers in an oil refinery belonged to fewer organizations—both in number and kind—went to fewer meetings, became officers of these organizations less frequently, and attended church services less often. Rotating shift workers spent a good deal more time around their houses than did the day workers.

An increase in shift work may also have some effect on health and on the ratio of men to women in some night shifts. At any rate, if shift work does become a necessity, the effect on society will be obvious. Most service facilities will have to be worked on shifts also. It may be that the churches will have three-shift ministers and the bowling alleys will be open continuously.

All in all, increased free time will stimulate questioning of our social system, certainly on such subjects as the distribution of wealth, work, and leisure. If these subjects do not spontaneously become public issues, the unions will see that they do. They will force management to think about how this wealth and productivity of machines should be distributed. Such questioning is already occurring.

Of course, it is possible that mankind may adjust to this freedom and that leisure may provide the opportunity to deal more significantly with what some feel are the most important tasks of life—care for home and family, service to the community, exploration, and the gratification of self.

If this adjustment does not come about spontaneously how do we go about finding a solution? Part of the solution of course, is economic, and it involves, as suggested before, distribution of wealth and work. But a good part of it involves the development of norms which will allow the individual to use a portion of his leisure, at least, as the ancient Greeks suggested, to contemplate, or, as that modern philosopher Harry Golden would say, to "enjoy, enjoy."

However, if we cannot get away from our Protestant Ethic, it may be necessary to have planned leisure. But before this could be done, a value system would have to be developed which would allow us to plan in an area that our mores now say belongs to the individual, and not to the group.

It would seem to me that increased leisure, in terms of free time, has to come. It could be devastating, but it need not be.

21 Edward H. Blakelock, "A Durkheimian Approach to Some Temporal Problems of Leisure," *Social Problems,* Summer, 1961, pp. 9–17.

IV

THE POLITY

Introduction

The impact of the computer on the political system has thus far been less dramatic than its effects on the economy. But the potential for change that is latent in current computer usage in the political sphere is at least as great, if not greater. While few computerized information systems for decision-making have yet been established, their introduction raises many thorny questions about the nature of political decisions in a democracy.

Public concern has already been generated in connection with two other political uses of the computer: public opinion sampling and the forecasting of election results. The advent of frequent and widespread polling has raised questions as to the nature of leadership in a democratic society. To what extent should the leader "follow" public opinion and to what extent should he be "above it"? This is a tension that has been endemic to the American political system, but recent advances in polling have helped to aggravate it.

Similarly, the attempt of politicians and political analysts to predict election results on the basis of such voting bloc factors as ethnicity, class, and region is not a new phenomenon. The computer programs set up for this purpose do essentially the same thing, though in a more sophisticated fashion; and they are not infallible. As one political forecaster has put it, "in the last analysis,...despite the fact that any given pattern may be simple, the question of which particular pattern will emerge in a given election campaign is still something of an imponderable.... There will always be margin for error in computer prognosis of public opinion.... For those still ethically troubled by the political advent of computer analysis, let me add that potential protection against the trivialization of the electoral process is the same as it has always been: an electorate that responds to complex issues in a complex way, a way that defies stereotypes and formulas—and the computers."[1]

[1] Robert P. Abelson, "Computers, Polls, & Public Opinion—Some Puzzles & Paradoxes," *Trans-Action,* 5 (September 1968), p. 27.

If the computer helps to simplify the process of predicting election results, its potential impact on decision-making is more ambiguous. By providing decision-makers with large bodies of information and manipulating this information in a variety of ways, the computer may help to simplify and improve the decision-making process. On the other hand, while "the computer, used wisely, increases the options from which the user can choose,... this increase in options comes at the cost of increased complexity;...Simulation models...allow the user to explore many more contingencies and probabilities than would be possible if the computer didn't exist. But it means that the decision-maker has much harder decisions to make because he *has* to consider much more information—it's there to *be* considered and much more sophisticated information at that."[2]

While computers may provide assistance in manipulating this complex information, the most difficult problems remain in the hands of the decision-makers. Questions of values and of conflicting objectives do not lend themselves readily to computer programming. Some commentators have expressed the fear that decision-makers may be seduced by the computer into ignoring, or giving insufficient weight to, these more intangible factors. Thus, "one programmer may value cheap, efficient roads and may ask the computer to provide him with specifications for such roads, whereas another may value expensive, beautiful roads. In this second instance, the values are less clear-cut and more difficult to measure: What is beautiful? How much should be spent for how much beauty? Not only are such parameters difficult to define and measure, they may in some instances be difficult to admit or recognize.... [There is a danger, therefore, that] complex decision systems involving human parameters will be broken down into routine segments which are more or less independent of human reaction, and that the combination will then be called a credible simulation of the total system. Such a danger has always existed in all categories of problem solution; however, with the advent of increasingly effective computers, the danger is becoming more seductive and more far-reaching in the scope of its influence."[3]

This danger should not obscure the equally important possibility that computer technology will be of substantial assistance to decision-makers in arriving at wiser and more informed policy decisions. But there remain other fears associated with the use of information technology for decision-making purposes: will there be undue invasions of privacy? Will political power come to rest almost exclusively in the hands of experts? Who will have access to the information stored in the computers? As decision-making in all spheres—in the public sector, the corporate sector, the universities, etc.—may come to depend increasingly upon access to the

---

2 Donald N. Michael, *The Unprepared Society: Planning For A Precarious Future* (New York: Basic Books, Inc., 1968), p. 48.
3 David L. Johnson and Arthur L. Kobler, "The Man-Computer Relationship," *Science,* 138 (November 23, 1962), pp. 875–876.

best available information, the question of who will have access to such information becomes a most important one.

The very existence of computers may help to generate more information. It has been pointed out, for example, that the volume of laws has been growing faster than the population, and that computers may be of help in making it easier to find the law. At the same time, however, there is a danger that by so doing, the computer may encourage the creation of new laws at a greater rate.[4] A second danger is that the independent lawyer with a small practice, his clients, and other laymen may not have ready access to computers for legal research. Some lawyers have expressed the concern that laymen must not be denied access to legal information stored in computers, just as they are not now denied access to law books.[5]

The problem of access to public information is a part of the larger political question about the rights and roles of the citizen in a democratic society that is equipped with sophisticated information technology. What will be the role of the expert and his relationship to both elected officials and the citizenry? It is still too early to detect the patterns that will emerge as the use of the technology becomes more widespread. But the outcomes will depend upon the interaction between the nature and demands of the technology and the values and social patterns of the political institutions.

The selection by Paul Armer reviews the current state of computer usage in government. He notes that although computers offer potential benefits in cost savings and in the improvement and expansion of services, most of the computers currently used by the government are in the defense and space sectors. Further use of computers, Armer maintains, will add to the existing pressures for consolidation of local government, although in the near future this will take the form of consolidation of files rather than consolidated authority. The rate of introduction of computers will be held back by unfamiliarity of top-echelon personnel with electronic data processing; shortage of qualified manpower; local and jurisdictional barriers; and fears about invasion of privacy.

Alan Westin places information systems technology in the context of the "basic elements which determine the public decision-making norms of a society." These are constitutional, technological, intellectual, ideological-value, and intra-organizational components, and the social agenda. Westin discusses the ways in which contemporary social changes in the United States are affecting these decision-making elements. He then examines the types of data banks that are currently being established in the public sector and looks at some of the political questions that they raise. "We could," Westin argues, "enter an era in which corporations, organizations, and political administrations would rise and fall according to their capacity to recognize, gain access to, and use rapidly the basic information pools of

---

4 See Reed C. Lawlor, "Computers, Law, and Society—Where Do We Go from Here?," *The Practical Lawyer,* 13 (March 1967), pp. 10–17.
5 See for example R. N. Freed, "Computer Law Searching: Problems for the Layman," *Datamation,* 13 (October 1967), pp. 38–41.

the 'data-rich' society." Therefore, "the largest political debate over the information systems" will concern the questions of "who shall have access to and control over this new technology" and "how will society adapt to increased knowledge of its previously well-concealed pathologies?" The influence of the "technocrats" who use the new decision-making technology "will be a function not only of their specialized training, but also of support they receive from a new political consensus. . . . It is the task of future empirical research to investigate the initial impact of information systems upon traditional political coalitions and the consequences of changing coalitions for the distribution of social resources."

Raymond A. Bauer examines the possibilities and limitations of information systems for use in social planning. He argues that "formal information systems are, by their nature, better suited to servicing relatively narrow repetitive problems than broad, unique ones." Ideally, society-wide information systems are to serve as aids in the detection of the state of affairs, the process of evaluation and diagnosis, and the selection of courses of action. But there are difficulties all along this line. The difficulty of evaluation, for example, results not only from value differences, but also from inadequate knowledge "as to what will be the eventual consequences of the state of affairs that has been detected. . . . [Here] . . . the adequacy of the model one has, and consensus on a model . . . become most relevant." If such consensus is lacking, it is difficult to know what to measure. The success of any such information system, Bauer maintains, will depend upon the quality of the inferences that are drawn from the data, the *ad hoc* research that is undertaken to establish causal connections and evaluate the effects of public programs, and feedback to detect the consequences of actions which cannot be adequately anticipated.

John Diebold explores the ramifications of new information technology for the State Department, noting that "the application of technology not only changes the method by which an operation is performed, but frequently changes *what* is performed. Just as businesses are now able through technology to provide entirely new services to ever-increasing numbers of people, so will the scope, conduct and substance of foreign affairs change as technology is applied. . . . As the horizons of factual ignorance and misinformation fade, the decision-maker will be presented with vast new areas of choice." Information technology will affect who makes the decisions and what the decisions are about.

The next two essays examine the issues involved in the potentials of the computer for invading individual privacy. Carl Kaysen, a member of the commission which put forth the proposal to establish a National Data Center, argues that "while the fears raised by critics have real content, the problems are neither entirely novel, nor beyond the range of control by adaptations of present governmental mechanisms. . . . The data center would supply to all users, inside and outside the government, frequency distributions, summaries, analyses, but never data on individuals or other single reporting units." Technical safeguards can be established against such abuses as illegal access or the "cracking" of information safes. A national data bank is needed, Kaysen maintains, to remedy the defects in the Federal statistical system, which is currently "too decentralized to function effectively and efficiently."

Donald N. Michael's essay takes a futuristic look at the impact of computer technology on individual privacy and freedom. Invasions of privacy, he notes, are frequent occurrences even in the absence of computers. Moreover, "it would not be surprising if, in the future, people were willing to exchange some freedom and privacy in one area for other social gains or for personal conveniences" (as for example in a large-scale federal attack on crime). The potential dangers of the computer arise because this technology makes it easier to collect meaningful data about persons from a diversity of sources. But in the future, Michael suggests, it may well be that "we shall have a new measure of privacy: that part of one's life which is defined as unimportant (or especially important) simply because the computers cannot deal with it." While computerized information about individuals will make it difficult for individuals to correct or provide interpretations for the records of their own past histories, centralization and computerization of private information may also decrease the risk of illegitimate leaks of information.

While the idea of using computers for purposes of legal research "has been terribly IN," Sally F. Dennis points out that "there are in motion actually only a few ventures in which the computer is used in some way to 'find the law.'" Her essay discusses the difficulties in using computers for legal purposes.

But if the computer has not yet been exploited to help solve legal problems, its existence has begun to generate a series of new legal issues. Stephen P. Sims discusses the nature of these issues in six areas of the law that are now or will in the future be affected by computer technology: antitrust, evidence, constitutional law, torts, the judiciary, and competition. He explores such possibilities as the use of the computer as expert witness and such problems as the locus of liability for computer products and services. By and large, issues of this sort have not yet been raised in the courts, but there is little doubt that they will be in the near future.

FOR FURTHER READING

Robert P. Abelson, "Computers, Polls, & Public Opinion—Some Puzzles & Paradoxes," *Trans-Action,* 5 (September 1968), pp. 20–27.

Raymond A. Bauer, ed., *Social Indicators* (Cambridge, Mass: M.I.T. Press, 1966).

James Schlesinger, "Systems Analysis and the Political Process," *Journal of Law and Economics,* 11 (October 1968), pp. 281–298.

Martin Shubik, "Information, Rationality, and Free Choice in a Future Democratic Society," *Daedalus,* 96 (Summer 1967), pp. 771–778.

Alan Westin, ed., *Information Technology in a Democracy* (Cambridge, Mass: Harvard University Press, 1970).

Aaron Wildavsky, "The Political Economy of Efficiency: Cost-Benefit Analysis, Systems Analysis, and Program Budgeting," *Public Administration Review,* 26 (December 1966), pp. 292–310.

Harold Wilensky, *Organizational Intelligence* (New York: Basic Books, 1967).

Computer Applications in Government

PAUL ARMER

Stanford University

THE PRESENT

Of the total number of computers installed in the United States today, about 10 percent are used by the Federal Government. Of these, about 7 percent are in the Department of Defense, with NASA and the AEC accounting for another 2 percent. Thus, nondefense and nonspace related computer activities of the Federal Government comprise only about 1 percent of the total computers installed.[1] The percentage of computers used by State and local governments is about 2.5 percent, up by a factor of about 5 since 1960.[2]

A vast potential exists in the utilization of computers in government, for much of its activity involves information processing. A study for the State of California by Lockheed Missiles & Space Co. estimates that the State's 100,000 employees spend 50 percent of their time processing information.[3]

As in industry, the first phase of utilization by government has involved the mechanization of existing manual systems or of punched-card procedures. Even though this phase has hardly begun, the next—integrated data processing—is underway and organizations have begun to examine their operations to an extent they never did before. This second phase is proceeding at such a pace that some organizations will undoubtedly by-pass the first.

Reprinted from Paul Armer, "Computer Aspects of Technological Change, Automation, And Economic Progress," in *The Outlook for Technological Change and Employment,* Appendix Volume I to *Technology and the American Economy.* Report of The National Commission on Technology, Automation, and Economic Progress (Washington, D.C.: Government Printing Office, 1966), pp. 220–223, by permission of The RAND Corporation.

[1] "The House of Representatives Passed the Amended Version of the Brooks Bill," *EDP Weekly,* Sept. 6, 1965, pp. 14–15.

[2] "EDP in State and Local Governments at Mid-Decade." *Automatic Data Processing Newsletter,* vol. IX, No. 25, May 10, 1965.

[3] *California Statewide Information System Study,* Final Report, Lockheed Missiles & Space Co., July 30, 1965.

Government organizations have conventionally organized files according to use; e.g., police, welfare, accounting, education, employment data. As a result, there is extensive duplication in collection and storage. Information collected by one group is often unknown to others who could make good use of it. Because of jurisdictional, mechanical, or procedural problems, data are not shared efficiently among various groups. Often information which would be very useful to one organization is not collected at all, even when another group could easily gather it in the normal course of operation.

The information with which government is concerned deals with people, organizations, real property, and personal property. Much of this information is common to many agencies; i.e., a person or a piece of property falls within the purview of different agencies for different purposes. Thus, the integrated data processing approach suggests that data files be organized by person and property and shared by all agencies rather than each agency having a file of its own. Shared files would not only reduce costly duplication but provide more comprehensive information than would otherwise be possible.

INTEGRATION ACROSS FUNCTIONS

An example of a file organized around property, and thus across several functions, is the data bank of Alexandria, Va., which contains information about the city organized into two master files. The street section file contains 120 items of information about the 3,518 blocks and intersections in the city; the parcel file contains 91 types of information about 20,000 parcels of land. Recently, within 1 week, six governmental managers made requests for information. Typical of these were:

> Survey all intersections in the city showing those where 5 or more accidents occurred in the past 3 months and give the characteristics of each such intersection.
>
> For two proposed urban renewal areas, analyze the density and location of welfare cases, minimum housing-code violations, health hazards, fires, mortalities, crimes, and arrests.

All six requests were fulfilled from the data bank that same week at a total cost of 3 hours of staff time and $67.50 for computer rental.[4]

INTEGRATION ACROSS POLITICAL BOUNDARIES

Data processing can also be integrated over political boundaries. Because of urban sprawl and the high mobility of our population, information on the same individual may appear in the police files of a dozen or so cities, often in different States. Similarly, because of overlapping jurisdictions, health or welfare data on the same individual may exist in many places.

The Police Information Network (called PIN) of a nine-county area around San Francisco Bay integrates such information through a centralized

4 "Small Cities and Time-Sharing," *Public Automation*, vol. 1, No. 1, June 1965.

electronic file of warrants of arrest. For example, a policeman who stops a motorist can radio the license plate number and name to headquarters where the information is typed into an inquiry terminal connected over a communications network to a computer. Within 2 minutes the policeman will know if any warrants are outstanding against the person in any of the 93 separate law-enforcement districts in the 9 counties and if the car is stolen or wanted in connection with another crime. The California Highway Patrol also operates a statewide stolen and wanted automobile system.

DUPLICATION

Some 14 million name records were in the files of just 12 of the agencies involved in PIN, although the total population of the 9 counties is about 4 million. Alameda County, also in the bay area, discovered in the process of developing an information system integrated across several functions that the county had 57 different files on people.

Duplication also exists between State and Federal organizations, their field offices, and local government counterparts. To indicate the possibilities for overlap, the United States has 50 States, 3,043 counties, 17,144 towns and townships, 17,997 municipalities, 34,678 school districts, and 18,323 special districts.[5] The number of separate agencies among these 91,236 areas apparently has never been recorded.

BENEFITS TO BE DERIVED FROM THE USE OF COMPUTERS IN GOVERNMENT

1. Cost savings should result from the elimination of duplication, from increased productivity in information processing, and from better utilization of resources through planning. More comprehensive, current, accessible, and accurate information should result in better decisionmaking.

2. Collection of more revenues should result through better audit, followup, and uniform application of complex rules.

3. Services to the public should be improved and expanded. This benefit is most important, although difficult to measure.

EXAMPLES OF THESE BENEFITS

1. Cost Savings

The Lockheed study[6] estimates that increased productivity would reduce personnel costs by between 1 and 5 percent—more than sufficient to pay for the system; in fact, at the upper end of the estimate, savings per

[5] Philip Meyer, "Too Many Governments Spoil Your Griping, *Standard Star,* New Rochelle, N.Y., June 4, 1965.
[6] *California Statewide Information System Study, op. cit.*

year are 10 times annual costs. This estimate would seem conservative, since it does not consider savings due to elimination of duplication.

St. Louis County saved $3 million in construction costs of a junior college by using a computer program which schedules classroom use. Classroom utilization was increased from the usual 30 to 40 percent to 80 percent, and fewer classrooms were needed.[7]

The cost effectiveness methods, so dependent on computers, that were introduced into the Department of Defense by Secretary Robert McNamara and his comptroller, Assistant Secretary of Defense Charles Hitch, could be used throughout the Federal Government and at State and local levels as well for evaluating various approaches to a problem.[8]

2. More Revenue

The Internal Revenue Service reports that in 1964 people reported $2 billion more in interest income than on 1963 individual income tax returns—a 28-percent increase. The IRS credits the use of computers and the public's awareness of what they can do for most of the increase, even though its computers processed individual returns in only one of seven regions last year. Collections from tax-delinquent businesses also increased when computers were installed.[9]

The use of computers can also increase the collection of fines from traffic warrants (it has been estimated that the PIN system will generate $2 million a year in fines that would otherwise go uncollected), and can help track down estranged fathers of children on welfare rolls, etc.

3. Improved and Expanded Services

A certain amount of unemployment is due to the time it takes for the unemployed to obtain information about appropriate job openings and for an employer to obtain information on qualified applicants. An information system designed to match the unemployed with job openings should hasten the reemployment process, and not only reduce the human misery of unemployment but also clerical costs and the costs of unemployment insurance.

The PIN system discussed previously is a very limited example of the application of EDP techniques to law enforcement. A much more ambitious system is under development for New York State, involving a state-wide centralized information utility integrating the complete files of over 3,600 law enforcement agencies, including police, prosecutors, criminal courts, and probation, parole, and correction offices. The first phase is scheduled to begin in 1967 with the complete system to be operational by 1969.[10]

7 "School Scheduling," *Public Automation,* vol. 1, No. 2, July 1965, p. 8.
8 "The Computer and the Pork Barrel," *Saturday Review,* Aug. 28, 1965, pp. 14–15.
9 "C&A Capital Report" *Computers and Automation,* July 1965, p. 35.
10 Ezra W. Geddes, Robert L. Ewich, and James F. McMurre, *Feasibility Report and Recommendations for a New York State Identification and Intelligence System,* System Development Corp., Nov. 1, 1963.

Not only should better information lead to better law enforcement, but hopefully with much more information about crime, we might gain some insights into the causes and the cures for crime and delinquency and thus map out a broad strategy of crime prevention.

Another new application of computers is in the legislative process. With pending legislation and proposed amendments and their status kept in a computer, each day the legislators could be brought up to date on where things stood at the close of the previous session. Such a scheme has been used in Ohio for appropriation bills. When better methods for processing language are available, proposed legislation could be compared with existing statutes for conflicts. However, for some time the best that such systems can do is to retrieve pertinent sections from present laws for comparisons by humans.

Another long-term and somewhat tenuous benefit to the individual citizen is that computers will tend to expose unsystematic administrative procedures, thus encouraging administration to become more uniform and less susceptible to political "pull." However, being systematic and objective can lead to inflexibility.

PERTINENT TECHNOLOGICAL DEVELOPMENTS

Information and information processing will become less expensive and more readily available to government as well as to industry. Computerized management information systems will also become important tools, and the development of large-scale mathematical models for simulating the economy of a region or of the entire Nation should contribute significantly to economic health in the 1970's.

Of great importance to law enforcement would be the development of a fingerprint-recognition device.

EFFECT ON EMPLOYMENT IN GOVERNMENT

The application of computers will have a significant impact on productivity in government because government activities involve a great deal of information processing, and because there exists so much duplication in collecting, storing, and retrieving data. But it is doubtful that there will be wholesale displacements of personnel; certainly the experience of the Federal Government to date would indicate otherwise. In a report to Congress in 1964,[11] the U.S. Civil Service Commission reported that in 10 agencies reviewed, only about a tenth of 1 percent of their personnel had been separated, reassigned, or declined reassignment because of the introduction of computers and automation. These agencies, when asked to forecast their

---

[11] *A Study of the Impact of Automation on Federal Employees,* prepared by the U.S. Civil Service Commission and referred to the Subcommittee on Census and Government Statistics of the Committee on Post Office and Civil Service, House of Representatives, 88th Cong., 2d sess., August 1964.

future personnel requirements, anticipated a decline of 5,000 in clerical positions but increases of over 10,000 in other jobs—chiefly those related to computer operation. These numbers are all of the order of a fraction of 1 percent of the Federal work force.

This should not be taken as an indication of insignificant productivity increases. Services have been increased, Parkinson's law has been at work, and "silent firings" (a reduction in the need to hire new employees) are the rule. The 1965 *Manpower Report of the President* anticipates that the number of Federal employees per 1,000 population will decline by about 2 percent between 1964 and 1968. It further states that electronic data processing has been a major factor in reduced growth in Federal employment, and has led to substantial changes in the composition of work forces.

As an example of an extreme increase in productivity, the report cites the Division of Disbursement of the Department of the Treasury where output per man-year increased by more than 300 percent while the dollar-cost per 1,000 checks and bonds was more than halved between 1949 and 1962.[12]

Though totals remain stable, they can mask significant changes. For example, the Internal Revenue Service's computerization plan calls for the centralization in 7 regional offices of processing tasks previously performed in 62 district offices. Although a greater number of new jobs will be created in the regional centers, some 5,000 routine office jobs in the district offices will be eliminated. Similarly, the optical character-recognition device to be installed at Social Security Administration headquarters will replace 140 keypunch machines and their operators.

THE COMPUTER AS A FORCE TOWARD CONSOLIDATION

The Government use of computers will add to the already great pressures for consolidation of local government. Computers may strengthen the position of counties with respect to cities, of States with respect to counties, and of the Federal Government with respect to States. The most common form of consolidation to be expected in the near future is exemplified by the PIN system where only *files* are consolidated and shared, but not authority.

FACTORS INFLUENCING THE RATE OF INTRODUCTION

Technological feasibility is hardly a problem. Economic feasibility is certainly at hand for many governmental activities not now computerized; and as computers become less expensive, many more applications will become practicable. Within a few years, computerization of almost all large govern-

---

12 *Manpower Report of the President and A Report on Manpower Requirements, Resources, Utilization, and Training,* U.S. Department of Labor, March 1965.

mental information processing tasks will be economically feasible, with better service at reduced costs the justification. However, the rate of introduction will be held back because of the following:

1. The unfamiliarity of top-echelon personnel, in general, with the principles of electronic data processing and with recent advances in management science will remain a substantial deterrent, although attempts are being made to rectify this situation.

2. There is a serious shortage of qualified personnel required to do staff studies and analysis. Government has difficulty in competing with industry for well-qualified computer professionals, who are in great demand and command high salaries. This, plus the situation discussed in (1) above, will contribute to a number of failures in applying information technology to government processes.

3. Authority is diffused as to what information is to be obtained, who is to have it, and how it is to be kept. Laws will often have to be changed to permit integration of files.

4. There is a lack of cohesiveness and coordination among organizations. For example, some State agencies, particularly those dealing in welfare, health, and roads, are more oriented to Washington than to their own State administration.

5. Parochialism inhibits integration across various levels of government, especially where organizational changes are necessary. In the face of such parochialism, the Lockheed report[13] proposed a federation of semiautonomous computer centers linked by an information central with information about where specific data are located. However, when problems become severe, political subdivisions will accept centralization: e.g., metropolitan water districts, transit districts, and police information networks.

6. The introduction of integrated files on individuals will be slowed by the not unfounded fear of a "1984 Big Brother" life with its invasion of individual privacy.

Despite these deterrents, computer use by State and local government increased about fivefold during the past 5 years. Computing power undoubtedly grew much faster—in the country as a whole it increased by a factor of about 30. In the coming decade, computer utilization in government will grow faster than in the rest of the economy. Most applications in the near future will be the computerization of existing manual procedures or of punched-card systems. Some integrated systems will appear, especially in law enforcement and particularly after the completion of pioneering applications like the New York State Identification and Intelligence System (scheduled for 1969). Once the decision is made to go ahead, it will take from 7 to 10 years to build the California Statewide Information System, which will involve almost complete integration across all functions and all geographical boundaries. (The decision to centralize geographically only information about data rather than the files themselves lessens the difficulties.) In contrast, the less ambitious New York system, while covering all the State, does not represent nearly as much functional integration.

---

[13] *California Statewide Information System Study, op. cit.*

Information Systems and Political Decision-Making

ALAN WESTIN

Columbia University

In his thoughtful paper on "The Impact of Information Technology on Organizational Control,"[1] Thomas Whisler of the University of Chicago defines information-technology as having three components: (1) *The computer* ("The engine that drives the technology. Without it, the other components become theoretically interesting but operationally unfeasible."); (2) *Telecommunications* ("the development of data networks...permitting intra- or inter-organizational communication between data input points, computers, and output points..."); and (3) Management science techniques. Whisler mentions "such things as Bayesian decision analysis, linear programming, and various models which fit the elements of management computational and decision problems."[1] As this is applied to government decision-making, I would stress the use of techniques such as systems analysis, operations research, the planning-programming-budgeting-systems (P.P.B.S.), modeling, simulations, social indicators, and forecasting.

To Whisler's three elements I would want to add a fourth—I call this the behavioral-quantitative persuasion, the belief that collecting large amounts of personal and interactional data and subjecting it to quantitative analysis is necessary to learn why people or groups behaved a certain way in the past, and to enable analysts of such data to construct reliable predictions of how people will behave in the future.

Reprinted from Alan Westin, ed. *Information Technology in a Democracy* (Cambridge, Mass.: Harvard University Press, 1970) © the President and Fellows of Harvard College by permission of the author and the Harvard University Press.

[1] In Charles A. Myers (ed.), *The Impact of Computers on Management* (Cambridge, Mass.: M.I.T. Press, 1967), 18.

Based on these four elements, I would define a public technological information system as one in which a government agency generates extensive data with computers and communication networks, analyzes these data by certain system-defining categories, and applies the output in a systematic manner to questions of public policy which concern that agency. . . .

Speaking in long-range historical terms, we can say that there are six basic elements which determine the public decision-making norms of a society:

1. A Constitutional Component. The national arrangements of institutional and political power will obviously set the requirements for consultations and consent between the executive authority and other governmental institutions in a political system. In a federal system, this will define which decisions "belong" primarily (or even exclusively) at the federal, state, or local governmental level, and in a constitutional democracy, there will be rules separating "public" and "private" decisions.

2. A Social Agenda. This will establish the policy areas that public agencies are charged with pursuing. The social agenda arises out of the value goals of the society and the clash among its contending group interests. At times of social change, the social agenda is dominated by the need of a society to respond to pressing socio-economic or technological problems. Of course, this agenda is often affected by external pressures on the state, particularly its economic and military relations in the international community.

3. A Technological Component. The state of the technological arts will define the operating capacities of the society to collect and communicate information and to subject it to orderly processing for decision-making purposes once it has been acquired. The current movement toward rapid reporting of events and computerized storage and processing of data pools represent a spectacular change in the technological basis for public decision-making.

4. An Intellectual Component. Here, I mean to group the prevailing knowledge and theoretical assumptions in the society about causal relations among man, nature, and society, especially the intellectual community's assumptions as to what can be done to affect such relations through human intervention. The moves in Western European and American society from scientific-rationalist to Enlightenment to Darwinian to welfare-state theories between the 17th and 20th centuries illustrate this intellectual force at work.

5. An Ideological-Value Component. This will define what each society considers right or wrong for rulers to attempt through their decisions, both as to the general ethical and political ends of governmental action and the governmental means through which such ends ought to be pursued.

6. Intra-Organizational Factors. So far, the factors noted have been largely external aspects. Yet a broad range of works by sociologists, political scientists, psychologists, and historians[2] inform us that organizations have goals and procedures that relate to the maintenance of the organization itself, and that public agencies are part of this phenomenon. Such organizations always have latent as well as manifest functions; they employ informal as well as formal patterns of decision-making, producing various "disorderly" and "irrational" behaviors in the decision-making process that reflect these elements. The difference between decision-making "norms" and "operating realities," between "scientifically-defined" and "organizationally-defined" notions of rationality, and between social and organization goals may give rise to conflict in the society. Thus, the realities of operations within public agencies and in their immediate operating environment represent a final element affecting decision-making patterns in every society.

If this way of viewing executive decision-making and its determining elements has merit as a way of organizing our analysis, it should be possible to look at any political system in time, map out its basic decision-making processes, and examine the factors producing these processes. It should also be possible to identify changes in the determining elements and to identify when such changes are producing qualitative shifts in the pattern. This is what I want to attempt in the next section.

CONTEMPORARY SOCIAL CHANGE AND AMERICAN DECISION-MAKING PATTERNS

It seems helpful to re-order these six factors to reflect the degree of change that is currently taking place in American society in each area. Thus, I will be moving from the most changing, technology, to the least changing, constitutionalism. Furthermore, let me stress before I begin that I am not presenting these observations as breathless discoveries; they seem to me well-known trends in society or points of analysis that have become close to conventional wisdom among intellectuals concerned with social change. It is their cumulative effect, and the momentum that is being built up by their cumulation, that is of interest to me here.

1. The technological factor of computers and communication networks is the most obviously dramatic in its novelty and effect on decision-making. If man's tools shape his problems and his ways of dealing with them,[3] then the very presence of these fantastically powerful storage and calculating machines and the equally important circuits of rapid collection and dissemination of data throughout the society have to exert a profound effect

2 Without taking the space to list here the relevant major works, I would note that Harold L. Wilensky, *Organizational Intelligence* (New York: Basic Books, 1967) is particularly relevant to my discussion of how "knowledge shapes policy" and the roles of information collectors and evaluators within public and private organizations.
3 See, for example, Emmanuel Mesthene, "Man and His Tools," unpublished paper, 1967, available through the Harvard Program on Technology and Society.

on decision-making concepts. Of course, the arrival of these new technological capacities does not automatically mean that they must or even should be used. But in such a private-oriented, de-centralized-government and technological-worshipping society as the United States, the technological "can" usually does mean "will," at least until critics of the new technology raise enough protests to bring the question of using the new technology to the level of public debate.

In this sense, it may well be that our "social problems" are not, in any objective sense, more serious in their creation of stresses and problems for our population than at many other earlier points in our national history, such as the move to industrialization between 1890 and 1920 or the depression era of 1929–39. But modern communications makes our crises painfully evident every day, sharpening group and public discontent. Moreover, the effect of the new technological tools on their supporters is to raise in their minds the possibility of dealing with the increasingly visible social problems in a way never possible before. What they look to is the possibility that current social reality might now be reproduced for study and public decision-making in all its *real* complexity, measuring ongoing changes with a precision never before possible, and evaluating the real effects of social interventions before they are made (or sufficiently early during interventions to adjust the policies in response to the "feedback.") But to do any of these things, the machines must be filled with data. And this can no longer be the "administratively-convenient" data of past eras—whatever could be collected and processed manually at reasonable costs of time and money without embarrassing agencies politically by its presence. Now, vast pools of increasingly more detailed and "personal" information from individuals, groups, and organizations is needed to provide a "true" general data base for analysis. As this trend increases, information and the technology to use it become a powerful potential resource in the society; we could enter an era in which corporations, organizations, and political administrations would rise and fall according to their capacity to recognize, gain access to, and use rapidly the basic information pools of the "data-rich" society.

2. Under changes in our intellectual outlook I would put primarily the growing efforts to reunify knowledge for decision-making. The increasing emphasis in social science on interdisciplinary language and behavioral research; the connections among scientists, engineers, and social scientists that are involved in efforts to apply systems-analysis and similar approaches to social problems as well as military, space, and resources areas; and the growing belief that the "knowledge communities" will be the dynamic element (and the power-brokers?) of the "post-industrial society" all represent powerful forces moving bodies of intellectuals toward the new information-technology, creating a new coalition-elite around "scientific" decision-making.

This movement is obviously accelerated by the nature of our present social agenda. This is not just the fact that many new goals of our society require information-technology approaches, as with our defense systems, biological and medical innovations, economic forecasting systems, etc. It is also the fact that the collapse of traditional social restraints (police control of Negroes and the poor) and of older "limited government" programs (in welfare, management of cities, etc.) and the accompanying shift in social values that has led to the creation of new large-scale governmental programs (medicare, education, anti-poverty) all demand greater information-collection and trend-analysis than our social agenda ever required in our history. Moreover, as the failure to deal adequately with problems of the American physical and human environment gives rise to mass protest movements, those elements in the nation which are directly threatened by such protest movements are pulled away to some extent from other concentrations: corporations from the development of new middle-class consumer products, universities from the pursuit of scholarly knowledge "for its own sake," etc. Social crisis focuses the attention and efforts of such institutions on programs of social reconstruction. The fact that the contract and tax policies of American government enable such institutions to make sizeable profits in the process (or to be well "reimbursed" for their efforts) helps corporations, universities, independent research organizations, and organized reform groups to see such service to society as also consistent with their organizational interests and political survival. Such are the happy links between interest and service in the post-industrial society.

3. The ideological components of our changing decision-making theory may seem at first to represent a cross-pressure to the growing trend toward reliance on information-technology. The movements of Negroes, the poor, and the student-left for greater participation in local and national decision-making and against continued "power-structure" programs would seem to run counter to technologically based decision-making by professional elites. Yet several factors seem to be minimizing this counterpressure, at least for the moment. For example, there seems to be growing agreement among the dominant white elites that large-scale planning and programming is essential to effective social action. This is illustrated nicely by the currently shared assumption between business leaders and liberals that white middle and lower class political attitudes are so hostile to adequate social programs for the poor and the black that a resort to "technological" rather than overtly political programs may be the only politically possible approach to certain large-program social interventions at present. There is also the widely held assumption of many systems planners that there need be no unbearable tension between technological planning and out-group participation; the good systems men will be quite able to program "full" participation by new groups into their designs for new social processes and institutions, as well as to supply the "necessary" recognitions and protections for individuality and privacy in their operations. Thus the rise of political demands for participation, equality, and individuality have had, at least up until now, the paradoxical effect not of derailing the supporters of information technology but of spurring them on to renewed efforts to save the society before it explodes under planlessness and inadequate administration.

4. Sociological factors in the organizational milieu are also moving American society toward new decision patterns. I would group several

developments under this heading. One is the current attraction of information technology elements to organizational leaders and the development of a bandwagon effect on public and private managers. ("We need a computer!" "What we need is systems-analysis of our operation!" "Can't we find out why people and things are really happening the way they are?") Another is the impetus given to these techniques by governmental subsidy of research in these areas and increasing government expenditures for in-house and outside systems-approaches to social problems. The reason I put this trend here is that it contributes to the growing interconnection in research and action between the public and private sectors, as well as among federal-state-and-local governmental authorities.

5. This means a growing trend toward functional and "informal" unity in our organizational life, a set of connections that *may* be able to overcome the last factor—the lack of significant change in our constitutional structure of federalized jurisdictions and separated powers during the present era. The inhibiting effect of this condition on the kind of national decision-making that the above intellectual, ideological, and social-agenda factors press for and our technological situation offers opportunities to pursue, represents a major restraint, along with budgetary limits caused by the Vietnam War, on the spread of information technology in the 1960's. This has led some believers in information technology to press for major government "re-organization" as a necessary condition for realizing the benefits of information technology. Because they feel it should come, many such advocates feel it surely will, though this is usually presented as a case for legislative and executive reform rather than for constitutional change.

So far, the discussion has stressed the ways in which many of the basic factors that determine a society's decision-making pattern seem to be receptive to experimentation with information technology. But it is not necessarily the case that social conditions that enhance experimentation are also favorable to the achievement of significant social *effects* by information technology in the near future.

For one thing, the degree of conflict and turmoil in American society over the Vietnam War, black-white relations, and institutional modernization may prevent information technology from obtaining large-scale funding, receiving cooperation from client groups and data populations, or being permitted to attempt innovative technological and organizational solutions. There may be some areas more than others in which information technology could move from experimentation to implementation even in an era of high political conflict and dissensus—such as medical care or employment systems—but these may well be limited to a few spheres.

If a new consensus of values and modernization of institutions does take place in the early 1970's, then information technology might come into a significant test period, especially if participative and civil liberty ingredients were solidly installed in the systems. One thing that seems to me doubtful, however, is that information technology systems will make any significant contribution to the possible emergence of a new political consensus. There is a body of opinion that sees technology as reducing areas of ideological conflict, by creating agreed upon spheres of "technical" or "professional" expertise. When problems are fundamentally political, however, with racial conflict heavily involved, I think it is asking more than information technology can deliver at this or any foreseeable stage to expect its systems to do any more than improve some program capacities in the accepted policy areas. . . .

THE DATABANK ROUTE TO TECHNOLOGICAL INFORMATION SYSTEMS

The most common classification of databanks, developed by Edgar Dunn[4] and popularized during the National Data Center debates of 1966–1967, distinguishes between "statistical" and "intelligence" systems. A statistical system is one organized to receive or collect data on individuals or groups in order to study systematic variations in the characteristics of groups. The purpose of the system is to conduct research and policy-planning studies. Though it requires identification of data by the individuals in the sample populations in order to associate new data with older holdings and to conduct longitudinal studies, data on individual persons are not the intended output of a statistical system. Furthermore, it is not intended to serve as a means of regulating or prosecuting individuals.

An intelligence system is one in which the data are deliberately organized into "person" files or dossiers to furnish reports about specific individuals. It is precisely to centralize data about individuals now held by scattered sources within government (and perhaps by private data holders as well) that the intelligence system is created. Intelligence systems can be used for research and policy studies, but their primary purposes are administrative (personnel management, licensing, payroll, etc.) ; regulatory (taxation, welfare, zoning, etc.) ; or punitive (police and law enforcement, national security, etc.).

This two-fold classification separates databanks according to their basic purpose and the directness of their regulatory effect on particular individuals. While these categories have value for considering the different policies that might be set for protecting privacy and providing due process in administering the two types of databanks, I think this is not a useful way to consider databanks in the decision-making process. "Intelligence" is a word that carries too many investigative, law enforcement, and loyalty-security connotations, and the databanks developing in education, transportation, welfare, or urban management ought not to be saddled with such an emotionally-charged label. In addition, "intelligence" is too gross

4 Edgar S. Dunn, Jr. "The Idea of a National Data Center and the Issue of Personal Privacy," *The American Statistician,* Feb. 21, 1967. See also Kenneth J. Dueker, "Data Sharing and Confidentiality in Urban and Regional Planning," a paper delivered at the Urban and Regional Information Systems Association Conference, Garden City, New York, September 8, 1967.

a category, and can be separated into several operating categories with greater profit to analysis.

In the typology that follows, I will describe five types of current databanks, based on a combination of their location within the organizational structure of the executive branch and the central purpose of the databank.

Type 1, the *Statistical Databank for Policy Studies,* is an agency usually attached to the political executive that collects information for research and policy-planning; it has no operating responsibilities to administer, regulate, or prosecute. It may be available to several or all agencies and departments for statistical services and special reports, or it may be used solely by the political executive (mayor, governor, etc.) or his principal administrative agency (budget bureau, controller's office, etc.).

An example of Type 1 is Detroit, Michigan's "Physical and Social Data Bank."[5] Though developed primarily to plan and study the city's urban renewal projects—Detroit spent $200 million in this area in the past decade—the databanks are available to any city department. As described recently in an article by two Detroit officials,

> The physical databank contains such information as the condition of the city's residential, commercial, and industrial buildings; property assessment figures; age characteristics of various structures; type of structure; the estimated costs of various kinds of physical treatment of residential structures (e.g., conservation, redevelopment and code enforcement); population characteristics and occupancy patterns; and many other kinds of data. The information in this bank is collected from census material, local surveys and studies, and our board of assessors' records (Detroit, for example, rated most of its residential areas, structure-by-structure, on a seven-point scale ranging from "sound" to "extremely dilapidated" in order to get an idea of the extent of residential blight in the city.)[6]

The managers of the physical databank report that "it has already served us well."

> "The blight ratings have been especially useful helping us to make decisions about such things as where clearance projects might be started, where rehabilitation might prove more effective, what our city's overall housing resource is, and so on. . . . . Moreover, since so much of our physical data is so readily at hand (and so easily manipulated), Detroit has enjoyed something of a head start in applying for federal grant-in-aid programs. Many of the grant applications require a great deal of physical information, in a number of different forms, and with our computerized data processing systems we have been able to come up with the necessary data in short order."[7]

The second step in the Detroit databank was to add, in 1966, a social data module to the physical data start. The social databank draws its statistics on a monthly basis from private service organizations as well as

[5] Harold Black and Edward Shaw, "Detroit's Data Banks," *Datamation* (March, 1967), 25–27.
[6] *Ibid.,* p. 26.
[7] *Ibid.,* p. 26.

city agencies, organized according to census tract. Data include "statistics on crime rates, welfare, births and deaths, school truancy and drop-out rates, the occurence of venereal diseases and tuberculosis, and other information."[8]

City officials report that the social databank has also "proved to be an immensely useful tool." It was used in writing the city's application for a federal Demonstration Cities Act grant; by the city board of education, which "can use the crime and truancy data to see if more emphasis might be put on school-oriented after-hours activities (such as athletics, special vocational training, teen club programs, and the like),"[9] and by the neighborhood rehabilitation program, which needs to know the "attitudes, motivations, and social characteristics" of the people with whom it works.

Viewing the databank development, Detroit officials state that "the age of computers has ushered in a new age of urban planning." Their description of progress was published in March of 1967; four months later, observers of the "new age of urban planning" in Detroit saw the city go up in flames, urban renewal projects and all.

Another example of a Type 1 databank is the "District of Columbia Real Property Databank." The basic unit of record is a lot or parcel, identified by address, lot and block number, census tract, police and voting precinct, school district, etc. and containing such information about the property as its size, land use, zoning, taxes, improvements, assessments, etc. . . .

Type 2, the *Executive Databank for General Administration* is usually attached to the political executive or to his central budgeting-and-control agency (city controller, federal Bureau of the Budget, etc.). This is the technologically-minded public administrator's dream, the "total management information system" that has all the operating agencies at a governmental level "on line" in "real time" with the data generated available for system-oriented decision-making. So far, the pioneers with this concept have been city and county governments.

In Alameda County, California, whose one million population includes Oakland and Berkeley, an administrative databank was begun in 1965 with the initiation of a "People Information System."[10] In March of 1967, this contained files on 200,000 individual county residents; when interrogated, the system will "print out what the computer knows" about John Jones when he applies for "social services like welfare, hospitals, health, and probation" or if he is the subject of an inquiry by any of the 93 law enforcement agencies in the Greater San Francisco Bay area. Alameda County plans to increase its present system to "embrace other county people files" and its long-range goal is to link an expanded "People Information System" with a "Property Information System" and an "Administrative Services System." . . .

In New Haven, Connecticut, a pilot project in collaboration with IBM

[8] *Ibid.,* p. 27.
[9] *Ibid.,* p. 27.
[10] Gordon Milliman, "Alameda County's 'People Information System'," *Datamation* (March 1967), 28–31.

has been started to place comprehensive record files on each city resident in a computer information system, develop communication networks for direct on-line registering of events and trends in city agencies, and to apply systems-analysis to the city as a governmental operation in order to develop better information for effective decision-making.[11]. . .

A detailed description of expectation and plans is found in the Systems Concept Report for the New Haven Project.[12] To provide "the kind of rational decision-making that contributes to effective urban government," the report states, each department and each level of city management need access to the proper information. Yet a study of present city data collection and use, the report states, shows serious problems of "comprehensiveness, accessibility, relatability, timeliness, and utilization." Under "utilization," the report stressed the need for adopting "advanced management science techniques" to make use of the improvements in information collection and flow that the city databank would provide. . . .

> Today, the complexity of the urban environment makes it all but impossible to evaluate the implications of a given decision. What, for example, would be the implications of making a vacant area of land a high school? A branch library? How much traffic would it generate? How would this traffic relate to existing road capacities? What types of families would be attracted to the area? What services would be required? What is the comparison of services for alternative courses of action? Many of these questions could be answered through the use of simulation models. A basic reason, however, why simulation has not been used by urban management to significant degree is not the applicability but rather the lack of quantitative data needed to develop the requisite interrelationships. It is believed that the creation of an Urban Management Information System would provide much of the required quantitative data.[13]

Perhaps the most self-conscious attempt to construct a technological information system—and our final example of Type 2—is the South Gate (California) Municipal Management Information System (SOGAMMIS), a collaborative effort of a "medium sized city," (58,000) and the Municipal Management Information System Research Project of the School of Public Administration of the University of Southern California. The project director, William H. Mitchel, formerly a U.S. Budget Bureau analyst and organizational consultant for Griffenhagen-Kroeger Inc., has written that SOGAMMIS "is a combination of systems theory, decision theory as it relates to resource allocation (e.g., program budgeting), and a dynamic conception of data storage."[14] As Mitchel explained these assumptions further:

---

[11] "New Haven Plans a Computer Pool," New York *Times,* March 29, 1967, p. 39. See also "Everything's Up-to-Date in Kansas City," *Journal of Data Management* (October 1967), 30–33.
[12] *IBM—City of New Haven: Joint Information System Study,* Report No. 1, System Concepts, December 1967.
[13] *Ibid.,* pp. 22–23.
[14] William H. Mitchel, "Tooling Computers to the Medium-Sized City," *Public Management,* Vol. XLIX, No, 3, (March 1967), 63–73.

The municipal experience with the databank approach clearly indicated that while it was a significant step in the right direction, it was severely limited as a conceptual tool to be used in the development of an operating information system.

Essentially, the databank concept is static. It involves the notion that a bank or master file of information is accumulated about a given city or urban area and stored in computer-accessible form. The city decision-makers can then draw on this information as needed. Computerized storage and retrieval arrangements insure that the desired information can be recovered rapidly and relatively inexpensively.

The databank file is updated periodically rather than as conditions actually change. The maintenance of a current file therefore constitutes a major problem, and is solved, if at all, by superimposing a data collection and reporting system upon existing operations.

The dynamic storage theory approach to data processing, however, provides that inputs are captured on a continuous basis and at the point and time of generation. The data sources therefore are the routine procedures required for the conduct of city business. Such data become—at the point of generation— part of the ever-changing data resource which can be made available at call to the full range of city activities without further demand on the operating departments whose activities initially generated the information. City planners, for example, can have available current information generated by and relevant to the day-to-day functioning of the police, fire, license, and public utility groups. Further, these computer processes are also an organic part of the routine operations of the several departments which direct benefits to these operations as well as to the larger, periodic, and non-dramatic decisions of top management and the council.

Using these basic notions as criteria against which to evaluate the information component of municipal activities, it was possible to evaluate, in general terms and in the specifics of South Gate operations, both the existing and hypothetical structure of city operations and determine how each could be adapted to the peculiarities of the computer.

SOGAMMIS thus is an effort, as its name implies, to think through and develop a computerized, total, integrated municipal management information and decision system, which considers both the historically sanctioned ways of doing business and how a city might operate when forced to change by current conditions and computer capability.[15]

Unlike many of the other experimental projects or administrative databanks moving toward technological information systems, SOGAMMIS presents an explicit discussion of the organizational and political problems involved in such information system ventures.[16]

Type 3, the *Independent Agency Databank,* is an agency formed to collect and store information from various departments and agencies, usually at more than one level of government, and to perform various information analysis and data-dissemination functions for the participating agencies. Like the statistical databank, the coordinating databank is not created to

---

15 *Ibid.,* pp. 66–67.
16 See *ibid.,* pp. 67–73.

operate substantive programs but to serve as an "information handler" for the "line agencies."

A leading example of Type 3 is the New York State Identification and Intelligence System (NYSIIS), developed by a gubernational study group in 1963 and created by legislative act in 1965 as an independent state agency to facilitate "information sharing" among the 3600 criminal justice agencies in New York State. . . .

To provide the data pool, NYSIIS is converting five million sets of fingerprints and more than a million criminal records from their present manual-file form to computer storage. It will use more than 70 million "active files" maintained by state agencies of criminal justice, and conduct searches of these files for participating agencies (such searches now average eight million a year). In its "Building Block One," NYSIIS has installed a facsimile network of transmit fingerprints and record-data throughout the state, is developing a new computer-assisted fingerprint classification system and a system for automatic scanning of vehicle license plates, and is creating an "organized crime intelligence capability" that will have NYSIIS intelligence analysts doing special studies of crime patterns and anti-crime strategies for use by state agencies. "In addition to this major intelligence module, NYSIIS plans to include as soon as possible, among others, capabilities for detailed criminal history, warrant-and-wanted notices, and other information relating to stolen vehicles, property marks, stolen property, laundry marks, handwriting, modus operandi, missing persons, voice prints, scientific and criminological research, and crime pattern analysis.[17]

California is developing a similar statewide criminal justice information system and already has state coordinating databanks in the motor vehicle[18] and educational fields.[19] Florida's state Department of Education has installed a coordinating data system called SPEDE (System for Processing Educational Data Electronically) which it describes as a "statewide total pupil information system."[20]

The fourth type of databank is the *Intra-Agency Databank* maintained within a particular department or agency in the executive branch. The United States Secret Service has been developing a computerized databank of potentially dangerous persons or groups who might attempt to assassinate the President. The Federal Bureau of Investigation has started a National

---

[17] Robert R. J. Gallati, "Computerized Information Against Crime," *State Government* (Spring 1967), p. 110.
[18] R. E. Montijo, Jr., "California DMV Goes On-Line," *Datamation* (May 1967), 31–36.
[19] A. Grossman, "The California Educational Information System," *Datamation* (March 1967), 32–37.
[20] L. Everett Yarbrough, "The Florida Project: A System for Processing Educational Data Electronically," *Journal of Education Data Processing*, Vol. 3, No. 2 (Spring 1966); *System for Processing Educational Data Electronically* (Tallahassee, Fla.: State Department of Education, 1966); *Activities Summary of Information Systems* (Tallahassee, Fla.: State Department of Education).

Crime Information Center to store data on wanted persons, stolen cars, and certain types of stolen property. Local and state law enforcement agencies feed in reports of crimes and can make inquiries about suspicious persons, vehicles, or property in their jurisdictions. The NCIC will be expanded to cover "additional criminal information" in the future, but the direction in which the system will grow has not been made public.

The California Department of Motor Vehicles has developed the Automated Management Information System (AMIS), which stores files on the individual's driver's license, vehicle registration, legal files of his history of moving violations or court convictions, accident reports, and other related motor vehicle data. This information is provided to drivers and to insurance companies for a nominal fee; it also creates instantaneous access to the information for management decision-making purposes. Included in the AMIS systems is an automated databank in Sacramento with a large storage capacity and an on-line system linking DMV field offices with courts, law enforcement agencies, and as many as 1,400 terminals.

A final example of a single agency databank is the Federal Manpower Information System.[21] This project, recommended by an advisory committee and currently in a planning stage, would provide consolidated personnel information on each federal employee for use by the United States Civil Service Commission for "manpower-management" purposes.

Type 5, the *Mixed, Public-Private Statistical Databank* is a system in which data from public agencies and private groups are given to a non-governmental organization which produces statistical studies of public programs. For example, the United Planning Organization, a private corporation financed by federal funds to conduct anti-poverty programs in the District of Columbia, plans to create a Social Databank to examine social problems affecting the poor and study the impact of D.C. anti-poverty programs, under a grant from the U.S. Department of Health, Education, and Welfare.[22]

The District of Columbia government has authorized release of approximately 81,000 individual records to the databank, primarily from education, police, and welfare agencies in the District. To protect the confidentiality of the data, UPO will create a trust, controlled by three independent trustees, who will oversee the databank to insure that no information about specific individuals is ever issued by the trust or released to government agencies. Mixed public-private databanks may well develop in the next few years in the medical field, where the creation of regional health information centers will depend on creating an "acceptable" agency outside

---

21 See *Modernizing the Management Information System for Federal Civilian Manpower,* Complete Report, Washington, D.C.: U.S. Civil Service Commission, Policy Development Division, Bureau of Policies and Standards, March 31, 1966.

22 "UPO Votes to Collect Wide Data on the Poor," Washington *Post,* January 17, 1968; Statement of Wiley Branton before the U.S. Senate Subcommittee on Administrative Practice and Procedure of the Committee on the Judiciary, February 6, 1968; Edward M. Brooks, "The Role of the Data Bank in UPO," Research Division, United Planning Organization, May 25, 1967; Third Draft, Trust Agreement between United Planning Organization and John Doe, January 19, 1968.

the regular governmental structure to receive and use health data from private as well as public sources.

As this brief catalog illustrates, various types of computerized databanks are appearing throughout the governmental and semi-governmental structure at every level of American government. At the same time, systems analysis procedures and other management science techniques are springing up across the governmental landscape, and serve as another path toward the development of technological-information systems. . . .

POLITICAL IMPLICATIONS

Information systems are intended to produce more efficient decision-making in the sense of reduced cost of operations and to increase the probability of "the best" policy alternatives being selected. Moreover, information systems are expected by their advocates to contribute to the strategies of governmental action by increasing the number and range of feasible alternative policies. . . .

In short, information systems promise to deliver a "meta-technology" or control technology to monitor the primary technologies (in production) and to create planned social change. Thus, the most interesting characteristic of information systems as foreseen by their advocates is that they are not merely tools to measure change but rather they are tools to guide social change.

The largest political debate over information systems arises directly from this "meta-technological" character of the systems: who shall have access to and control over this new technology, how will society adapt to increased knowledge of its previously well-concealed pathologies? How compatible are centralized information and policy formation procedures with broadly based democratic participation? How shall information and knowledge resources be distributed? . . .

One of the consequences of establishing an information system is that a large body of facts about the social order is gathered under a single roof and prepared for analysis by agency personnel and agency consultants. As pressure mounts from both the legislature and the political executive for initial payoffs from the large social investment made in the system, the information system must determine the political costs involved in releasing findings uncomplimentary to the agency policy, executive approach or party policy in that area. Much of the information collected by existent systems can have the effect of destroying entrenched myths concerning the state of society and the effectiveness of past governmental policy. An important question thus concerns the behavior of the political executive in this crucial phase: does the political executive attempt to edit findings of the information system, limit publication, or failing this, attempt to relegate the system to the status of a study group? During this period the status of traditional agencies may rise in terms of their influence with the political executive, given their ability to handle policy and information "through channels."

A second public area of impact is the Collectivity—the clients, regulated groups, or beneficiaries of each information system. The outputs of the information system will obviously have policy implications for the Collectivities under the jurisdiction of these agencies. The Collectivity at this stage, if it does not agree with the findings and analysis of the information system, may organize to either discredit these findings or establish their own information system to countervail the government system. A second possible tactic of the Collectivity is to attempt to build general political support in the electorate for its cause vis-à-vis the information system, and thus indicate to the political executive that the policy implications of the information system's work are politically too costly.

A third area of impact is the Courts: how will they consider the question of "reasonable government action" when a dispute arises between an information system agency and its clients? In a more traditional setting broadly educated advocates could argue their cases in court and expect the judge to adjudicate their case by reference to precedent. In the case of information systems the Courts may well have to restructure themselves through specialized training of judges or through the attachment of information system experts to their staffs.

Finally, the information systems will have a number of consequences for the relative state of consensus and distribution of power within the society as a whole. Information systems may generate knowledge which undercuts traditional governmental justifications for inaction, and this same knowledge can be used by dissenting groups to intensify political pressure for a redistribution of societal resources. A number of the ideological conceptions of a new democratic society may come under attack in this process as new knowledge about society conflicts with prevailing political myths. The prospect of democratically delegated authority being subverted by systems technologists, and the implied reduction of democratic politics to a doctor-patient relationship between the state and private groups, is already a familiar Sunday magazine caricature. More serious speculation on the growth of a "technostructure" and a societal group of "technocrats" would suggest that their influence will be a function not only of their specialized training but also of support they receive from a new political consensus base. This is a speculative hypothesis. It is the task of future empirical research to investigate the initial impact of information systems upon traditional political coalitions and the consequences of changing coalitions for the distribution of societal resources.[23]

---

23 The section on political implications [pp. 143–144] is based on a collaboration between the author and his research assistant, Kenneth Laudon, Lecturer in Sociology at Columbia University.

Social Planning

RAYMOND A. BAUER

Harvard University

The concern over measurement of social phenomena...must be viewed as part of a larger concern with the planning and control of society which has developed in the United States in the past decade, a matter in which we have lagged somewhat behind other Western countries. The shift in the American attitude toward planning might be traced to very many sources. Probably the key factor was our success in our management of the economy in the period since World War II. But perhaps almost as important are our changes in affluence and in the nature of the society itself....

A modern version of planning and control places a high premium on early detection of the consequences of one's actions, with a consequent adjustment of one's plan. Detection of these consequences may cause one to take different steps toward the goal, and to alter that goal or goals. To a large extent, this view of planning and control is influenced by the cybernetic model of electrical engineering, which stresses the importance of feedback to correct errors resulting from one's actions. While the cybernetic model has undoubtedly had a strong impact on our view of planning and control, it does not *per se* provide for the reassessment of one's goals—namely, for the correction of one's course toward an established goal. Probably the most profound contribution of cybernetics to our thinking is the establishment of error as a systematic inevitable feature of all action.

The notion of adjustment of goals comes from an approach to planning and control that stresses the plurality of future possible states and consequences of one's actions.... One uses whatever information and stimulants to one's imagination that are available to anticipate that range of future states of reasonable possibility and importance that might flow from one's

Reprinted from Raymond A. Bauer, "Societal Feedback," *The Annals of The American Academy of Political and Social Science,* 373 (September 1967), pp. 181–188, by permission of the author and The American Academy of Political and Social Science.

actions. (De Jouvenel calls these "conjectured" future states *futuribles*.)[1] One then decides which of these future states one wants to make most probable. (A highly desirable state may appear difficult to attain, and therefore a less desired state may be aimed at.) Having chosen such a target future state, one then devises and inaugurates a course of action aimed at increasing the probability of its occurrence. Having inaugurated that course of action, one takes readings of the consequences of those actions, reassessing the probable future states and the probability of their being attained, and making the adjustments of action and goals referred to above.

Clearly, in this version of how one should plan and act, a high premium is placed on accurate and rapid information about the existing state of the system on which one is acting and on one's ability to relate those states of the system to actions already completed. It could be contended both that no model of planning and control ever devalued information and that the model, as I have sketched it, is one which is seldom if ever used. All I would contend is that to the extent that this model is approximated, feedback becomes a more crucial element. . . .

FORMAL INFORMATION SYSTEMS

In recent years, "information systems" have come into considerable vogue as a management tool in business. But, in practice, at the time this is being written, the number of sophisticated information systems in operation is apparently very small. Despite this, the logic of their use is so persuasive that this use is bound to spread. An information system is no more than a formal set of procedures for gathering, storing, retrieving, and reporting data relevant to the decisions and actions that must be taken in an organization. In their prototypic fashion, information systems may actually take over some of the decisions and actions ordinarily done by people, by feeding the relevant data directly into a machine or other mechanism for action. Thus, sales or shipping records may be fed into an inventory system which will automatically order additional production to fill the inventory or order shipments to retail outlets to replace items which have been sold.

Such systems for the control of routine operations have been in existence for more than a decade. They have applied, however, *only* to such situations in which decisions and actions can be routinized and generally where optimum courses of action can be calculated. Of more interest to us, and of more recent origins, are information systems which are designed to serve those situations in which human judgment is essential. This latter class of situations is a large one. And while many of the decisions and actions in which human judgment is presently involved may become routinized in the future, there will always be some irreducible number of situations for which this is neither possible nor desirable. These are situations in which

[1] Cf. Bertrand de Jouvenel, *The Art of Conjecture* (New York: Basic Books, 1967).

the decision is unstructured, which is to say that the problem with which one is confronted demands the invention of new courses of action and/or demands trade-offs of values. A structured decision is one in which, for all practical purposes, the full range of actions is known; the consequences of the actions are known with substantial accuracy; and there is agreement on the value to be maximized.

Quite obviously, there are very few important issues in the public arena which involve structured decisions of this sort. At a minimum, there is no clear agreement on the weight to put on the various values involved or on whose interests are to be preferred over others, and usually there is the need for the invention of some unique course of action that will reconcile a sufficient coalition of interests. The logic of what a formal information system does for a person in such a situation is not new. It is substantially what the military have done for centuries in presenting generals with up-to-date position statements, displaying, in as appropriate a fashion as possible, information on his own and his enemy's situation. Furthermore, war games and simulated war games, such as chess, gave him a basis for anticipating the probable consequences of the various courses of action open to him.

What a modern computer-based information system does is to make available more information, more rapidly, and in many forms. In its most developed form it includes (1) ongoing data series such as "social indicators"; (2) stored data that an executive may want to use, for example, the demographic characteristics of the population a hospital serves, past records of illness in various segments of the population, and the results of experiments which have been carried out; and (3) a simulation model of the system on which the executive wishes to act. This model might, for example, be a model of the market for toothpaste, which will tell him the probable reactions of consumers and of competitors if he does something such as change the price of his product or the amount of his advertising.

We have, of course, precedents for such models in the public arena. The economic models used for control of the economy in the past two decades actually predated the computer-based simulation models used in business. The economists have, of course, updated their own models as the state of the art advanced.

Granting that there are earlier precedents or approximations for each of its components, the image of a modern, sophisticated information system *suggests* the possibility of planning and control of a sort that would have been impossible in the past. I have not mentioned yet, for example, the notion of a "real time" system which, for practical purposes, presents a man with a picture of the state of the world in which he is interested, with no time lag. Thus, sales of a company's products from wholesale houses, or potentially even from retail outlets, can be recorded directly in a computer as they are made so that a sales manager could know within a matter of minutes the precise state of sales of his product on a company-wide basis, by region, by city, or in any combination he chose. He could also know the state of his own organization, if he wanted to: the rate of production of various products, cash available, or, if he cares, the health of the labor force or the existence of unusual skills. Whether this is a good thing or not can be discussed separately.

The image referred to above is, then, that of a man with virtually instantaneous feedback of many of the consequences of his actions or, at least, instantaneous feedback of those things which he has chosen to measure. (There will be an organizational lag between his own actions and the impact of the organization's actions on the environment; for example, it may take many months for a program of inoculation to be organized and executed, and to have its effect on incidence of a disease.) In any event, he has a very up-to-date and complete picture of the world in which he operates and of the state of his own organization. Furthermore, his ability to anticipate the consequences of his possible future actions is improved by trial runs on a simulation model.

An information system of this sort is the logical tool for the type of planning and control which I described above. It in no way changes the rationale. The use of a simulation model to make trial runs is no more than a refinement of the process of "anticipation," but does not reduce the need for feedback. Since simulation models are based on historical data and are, of necessity, simplifications of the real world, they can only reduce, but never eliminate, the errors which will inevitably result from any course of action.

Systems of this sort have, of course, their own difficulties. For example, a person whose feedback system is "perfect" may react too rapidly to changes in reactions to his actions and cause great damage. The tendency of executives to overreact to information is a real and serious one. Also simulation models must, of necessity, make assumptions to bridge our gaps in information and in our understanding of the state of the world, and some of these assumptions turn out to produce seriously erroneous answers at times. And, of course, the information system does not tell one what courses of action are conceivable. Man himself must conceive them.[2]

Whatever its limitations, a formal, computer-based information system is a good prototype of the sort of system one might want for planning and control. It is also a good point of departure for consideration of the limitations of a societal informational system for planning and control in the public arena.

We may now move from this idealized world to the real one.

---

[2] Here, as at other places in this essay, the knowledgeable reader will be searching for qualifications. It is not true that an information system cannot be built that will generate possible courses of action. However, this development will generally have to wait until our understanding of human thought processes is better developed. Work on programming computers to play chess has shown that if one were to use the computer's tremendous calculational capacities to have it consider all logical possibilities, it would play an untenably long chess game. Its human opponents would have died of old age. Chess masters have heuristic devices—decision rules which simplify such enormous complexity.

LIMITATIONS OF FORMAL INFORMATION SYSTEMS

This section is devoted, in part, to an explication of the inherent limitations of formal information systems and, in part, to the extent to which a societal information system cannot, at this point in time, achieve the degree of sophistication that seems possible for the informational systems of single-purpose institutions such as the military, health systems, or business firms.

The primary limitation of any information system is that it is finite. But on the other hand, if it were not finite, it would be impossible to use, even if cost and effort were no consideration. The goodness of an information system is dependent on the extent to which it is efficient in serving the informational needs of specifiable individuals in an organization. Ordinarily, the development of an effective information system begins with an analysis of the decisions made by various persons in the organization, their preferences in the use of information, and the like. The purpose of this analysis is, in the first instance, to make sure that these individuals will get the information they need and want; but, in the second instance and equally important, that they will not get information that they do not need or want. The need to sort out unrequired information not only produces inefficiency but may even have the worse result of tempting a person to meddle in organizational affairs which are "none of his business." For example, the chief executive of a business firm who gets daily reports on sales in specific cities may begin interfering in tactical maneuvers that are properly the affair of the regional marketing director.

We may begin to differentiate a societal information system from that designed to serve a single institution by noting the multiplicity of users of the societal system. Not only must a vastly larger number of *actors* be served, but a societal system must also serve a vast array of *evaluators*: the public, the Congress, the press, and future historians, for example. It must be aimed at comparability over a very long time span. Hence, its data series cannot be closely designed to serve one clientele.

The most efficient information systems must, of course, contain some surplus information. If the flow of information to a man were restricted to only that which he was *certain* to need, it would inevitably miss information which probably would be useful to him. The amount of potentially surplus information to include is a difficult matter to decide, but the proper answer lies somewhere between "everything" and the certainly useful. This decision obviously can be made more sensibly for repetitive types of problems. The immediate corollary of this is that formal information systems are less useful for broad strategic problems than for routine ones.

Strategic or policy decisions are, by definition, responses to major opportunities or threats which could not entirely be anticipated. The first sign that a major threat or opportunity exists *may* manifest itself in some sharp discontinuity in regular data gathered for routine purposes, or in some shift in the relationship of two or more data series. One of the standard features of formal systems is "exception-reporting," which notifies the relevant person that some one indicator has deviated beyond some pre-established limit. Generally speaking, however, such changes of pattern will indicate merely that "something is happening," and thereby serve as a stimulus to search for pertinent data. It is just as likely that the signal for a strategic or policy decision will come from some source entirely outside the formal information system—a newspaper story, a conversation with an informed person, or some other. This matter of how strategic information is sought out will be dealt with in more detail later. At this point, all I wish to establish is that formal information systems are, by their nature, better suited to servicing relatively narrow repetitive problems than broad, unique ones. They can indeed be very valuable for servicing broad, unique policy problems, but such problems demand additional activities which will be discussed below.

We can get additional perspective on the limitations of a societal information system if we review the broad functions for which one would expect an information system to be useful. (Note that I did not say "serve." This was to avoid the implication that an information system could, even under ideal circumstances, be expected to do the entire job.) These broad functions are: (1) detection of the state of affairs, (2) evaluation, (3) diagnosis, and (4) guide to action.

Social indicators, like any other series of trend data, are *per se* means for detecting the state of affairs.

For some matters, the job of evaluation may be virtually automatic in that there is a strong consensus that some states of affairs (for example, health) are "good," and others are "bad" (for example, sickness). However, in the majority of instances, evaluation depends on the system model one holds, whether that model be explicit or implicit. For example, a setback in educational opportunity for Negroes might be viewed favorably if one thought that revelation of the situation would create such a feeling of indignation that it would create unprecedented support for increasing opportunities for Negroes. If the reader thinks that this type of reasoning is far-fetched, I would point out that it rests on logically the same type of process as that behind the frequently used argument that the pace of Negro progress should be contained for fear that the white backlash would more than offset the gains.

What is important, however, is that, in many instances, evaluation is not self-evident. And the difficulties of evaluation do not rest solely on personal differences in taste or values, but on empirical questions as to what will be the eventual consequences of the state of affairs that has been detected. It is here that the adequacy of the model one has, and consensus on a model among parties, becomes most relevant. It is gratuitous to say that, whatever difficulties one may have in developing a model of a market for a business firm, the difficulties of developing one for the United States social system, or even that of a metropolitan area, are infinitely greater.

The absence of consensus on a model, of course, poses problems at the very point of selection of what indicators to measure. What one has to look to is consensus on the indicators independently of consensus on a model of the society. For example, most of the relevant parties can agree that unemployment is something that we want to have information on, even though they do not agree as to what conclusions they will draw from the data.

The steps from detection, to evaluation, to diagnosis—let alone to prescription for action—are usually taken too blithely. Few, if any, of the social indicators proposed will tell us *automatically* what caused a given state of affairs. Accurate data may, however, spoil one of our favorite sports of invoking preferred explanations of dubiously existent phenomena, for example, blaming an "increasing" crime rate on the collapse of our moral fiber. Actual causal relations must be inferred, with the gaps of inference being narrowed by *ad hoc* research designed to establish the linkage. We may infer, for example, that the white backlash has been caused by some combination of the advances of the Negro community, anxiety over threats to their own position on the part of whites, moral indignation, guilt, or what have you. Research directed to this issue can, with varying adequacy, sort out these plausible causes.

Any information system requires provision for *ad hoc* research directed toward two ends: (1) diagnosis, for example, exploration of the causal origin of trends reflected in regular social indicators or of any other observable important social changes, and (2) measurement of the impact of discrete events whose impact may be expected to be reflected in regular data series only partially, indirectly, or with some delay.[3] Included in the latter category is the study of unexpected events such as a presidential assassination but, perhaps more importantly, as a regular matter, program evaluation. The evaluation of public programs demands rapid feedback, and it also demands measures of phenomena that one would not ordinarily think of including in regular indicator series. We may take as an example Project Headstart of the current poverty program. If it achieves its goal of increasing the opportunities of underprivileged children, this ultimate objective will be detectable only some years from now in the educational and occupational performances of such groups as Negroes. The immediate traceable impact may be found in such things as the degree of enthusiasm for the program on the part of parents and children (this at least will ensure their continued participation) and some improvement in the motivation and social skills of the children, which may be the preface to an improvement in learning ability and in learning itself and *subsequently* to improved educational and occupational performance.

While such measures are necessary in any information system, they play an increased role in a broad-gauged, multi-purpose system of social indicators. Or to put the matter in reverse, the more closely a system is tailored to a limited set of objectives, the higher the proportion of relevant effects which will be included in regular data series and the fewer will be the

---

[3] Cf. Albert D. Biderman, "Anticipatory Studies and Stand-by Research Capabilities," in Bauer (ed.), *Social Indicators* (Cambridge, Mass.: M.I.T. Press, 1966).

steps of inference required to establish causation. This is probably a tautology, but it underscores the general need for such *ad hoc* research in a societal information system.

Finally, any presently constructed societal information system will suffer from the *relative* lack of models with which to test out the consequences of possible courses of action. This lack is, of course, relative. Economists can test out the probable consequences of economic policies. One can also predict that a frontal attack on infectious diseases will almost certainly result in an increase in such diseases as heart trouble and cancer. On the whole, however, we lack dynamic models of most of the areas of our society which will help us to "anticipate" the consequences of programs we might introduce. This circumstance dictates the relative importance of feedback, of being able to detect these consequences as rapidly as possible after their occurrence.

In sum, formal information systems, like all of man's creations, have inherent limitations, and these inherent limitations get accentuated in anything we might contemplate in the way of a societal information system. Information systems are most adequate for handling repetitive, routinized problems and to the extent that the system is tailored closely to the needs of a limited range of problems. While societal information systems are useful for routine operating purposes, they are most pertinent for the handling of broader policy problems of an unprogrammed nature. Furthermore, a societal information system applies not only to a wide range of problems, but also to a variety of interests. It is not merely a tool for action for those who must devise and carry out actions (as in a business firm or health system), but also for those people who evaluate such actions. And a societal information system must be designed with a relatively poor model of the society.

The result of all these conditions is that the parameters to be measured must be selected only on the basis of a general agreement that they are "important," even though their importance cannot be justified *via* reference to an agreed-upon model. Furthermore, measures of such parameters must be, in general, oriented toward multiple usage. On the whole, this means, in turn, that the problems of evaluation and of diagnosis and of drawing inferences for action are based on complex inferences. Such a system will be relatively unusually dependent on the quality of the inferences drawn, on *ad hoc* research to establish causal connections and evaluate the effects of discrete events such as public programs, and on feedback to detect the consequences of actions which can only inadequately be anticipated.

JOHN DIEBOLD

The Diebold Group, Inc.

Two recent Presidential directives provide the framework for testing the application of the newest tools of information technology to the conduct of foreign affairs. If such tools are effectively applied and gain wider acceptance they could radically affect the management and even the substance of international relations.

On October 12, 1965, the President "directed the introduction of an integrated programming-planning-budgeting system [P.P.B.S.] in the executive branch," including the State Department. The system is a management method for measuring the effectiveness of expenditures in reaching program goals and had marked success when introduced by Secretary McNamara in the Defense Department. In implementing this system within the Defense Department there has been wide use of computer technology. Similar systems and technology are now being proposed and tested for the needs of the State Department.

The second directive was issued on March 4, 1966, when the President "directed the Secretary of State...to assume authority and responsibility for the overall direction, coördination and supervision of interdepartmental activities of the United States Government overseas." Within certain limitations, the Secretary now has the charter to become the manager of our foreign affairs rather than merely the coördinator.

The success with which the Secretary manages the State Department will depend to a major extent on his ability to meet its requirements for information and communications. These are now so complex that the question is no longer whether technology should be applied to meet them, but how. The success of such technology within the Department depends critically on three factors: (1) sound analysis at the highest level of the

Reprinted from John Diebold, "Computers, Program Management and Foreign Affairs," *Foreign Affairs*, 45 (October 1966), pp. 125–134, Copyright © by the Council on Foreign Relations, Inc. Used by permission of *Foreign Affairs*.

information needs of the Department; (2) the effective application of information technology to these needs, rather than simply the mechanization of the current inadequate information systems; and (3) the communication of the information thus collected to those who need and must act upon it.

To those who conduct our foreign affairs, as to the manager of a private enterprise, information technology poses not only questions of application but also challenges of change. For the application of technology not only changes the method by which an operation is performed, but frequently changes *what* is performed. Just as businesses are now able through technology to provide entirely new services to ever-increasing numbers of people, so will the scope, conduct and substance of foreign affairs change as technology is applied. Let me emphasize that what I foresee represents no panacea, no automated Foreign Service. My purpose here is to underline the fact that these new technologies raise major questions and require the most thoughtful planning.

The choices of instruments for decision and action are widening. The old obstacles to judgment and service are receding and are in the process of taking on new and, at this time, unpredictable shapes. It is my judgment, however, that as the new technology becomes applied to foreign affairs, reliance on personal judgment and personal and national moral standards will increase—not decrease. As the horizons of factual ignorance and misinformation fade, the decision-maker will be presented with vast new areas of choice.

If, for example, information systems are perfected by technology, what will be the role of the Ambassador? He could have available instantly all of the information and analysis available to the Secretary of State but might still lack the latter's overall view of national priorities and interests. Two or three hundred years ago, when it required days, weeks, months or, in some cases, years for an Ambassador to reach his assignment or to communicate with his sovereign, he was indeed plenipotentiary. There was no choice. He knew more than anyone at home about conditions in his assigned country, and orders regarding the most fundamental and long-term actions could not, in most cases, reach him in time to be relevant. Over the past one hundred years, with the coming of the telegraph, the wireless, the express train and the jet, the role of the Ambassador has diminished, at least in terms of his power to act. At home the number of people who know as much about his mission as he does has increased. As a matter of fact, the Ambassador's home office has at its disposal sources of information and analytical talents to which he has no access.

Now, however, the situation is changing again. If we so decide, the Ambassador will be able to have all the information relevant to his assignment. He could once more be designated in fact plenipotentiary if this were the wish of his superiors. On the other hand, as his home office will be able to be in even closer touch with his mission than before, the need to rely on the judgment of the man-on-the-spot could diminish even further. When the leaders of nations can confer for hours, face to face, on closed-circuit television, will the Ambassador's role become even more limited to that of an information-gatherer, pulse-taker and "holder of hands?" It is interesting to speculate on the kind of summitry we will have when such technologies really become effective.

Thus, the areas of choice between effective courses of action widen. Who makes the decisions? Who is the instrument of response? Other examples, perhaps more portentous in nature, will appear later. But even in this case of the Ambassador, the implications for foreign-affairs management are not to be dismissed lightly. If some sort of middle course is chosen, let us say by making the power of the Ambassador dependent on the sensitivity of his post or on his personal abilities, serious consequences to the prestige of our envoys could result. The fact that this problem has been developing for some time does not diminish its implications for the future. For, as the distance between alternative policies lengthens, deliberate or unconscious inconsistencies become both more obvious and more fraught with consequence.

II

Sir Winston Churchill, in discussing the process of making strategic wartime decisions, wrote: "Success depends on sound deductions from a mass of intelligence, often specialized and highly technical, on every aspect of the enemy's national life, and much of this information has to be gathered in peace-time." How much simpler the decision-making process might have been for him had it been possible then, as it is becoming increasingly so now, to centralize such information technologically.

In the State Department in Washington, some 2,000 telegrams are processed every day and an average of 70 copies is made of each. The resultant 140,000 pieces of paper daily are filed both centrally and in various user files. The Central Foreign Policy file alone grows at the rate of 600 cubic feet (400 file drawers) a year. The Intelligence staff has 200 professional employees who read and try to analyze some 100,000 documents monthly. Most of this information is filed to meet the personal requirements of those in charge of various bureaus and offices. Its existence is not known or useful to others. Senior officers must wade through stacks of telegrams and airgrams to get a few bits of significant information. The new or most important information is mixed with the old or trivial. In an emergency situation the central filing system is ignored almost entirely and a crisis team of experts on that particular situation or country is called together to offer its analysis and advice.

The problem of information-flow and use has been recognized in the foreign-affairs community of our government. Since 1946, some 363 projects and studies in information management have been undertaken, 172 of them in the State Department and 167 in the various successor agencies concerned with foreign economic assistance.

In one informational area of foreign affairs, technological solutions are being vigorously tested—collating information about the State Department's resources. A Foreign Affairs Programming System (F.A.P.S.) was established with the objective of bringing together all the strands of United States activities and resources abroad, country by country, to give both the Secretary of State and the Ambassadors a coherent instrument of command and control. Now elements of the F.A.P.S. are being refined to serve the P.P.B.S. being introduced at the President's direction.

In the spring of this year, Dr. Charles Hitch, formerly with the RAND Corporation and then, as Comptroller of the Department of Defense, architect of the programming system introduced by Secretary McNamara, was designated as the chairman of a newly appointed advisory group on foreign affairs planning-programming-budgeting. This advisory group is charged with developing a P.P.B.S. for the State Department. The Stanford Research Institute, State Department personnel, the Bureau of the Budget and the P.P.B.S. personnel of other agencies will work closely with the advisory group. It is expected that by fiscal 1969 a full P.P.B.S. cycle can be developed for Latin America—the first test region.

But major problems remain. Richard Barrett, Director of the Office of Management Planning in the State Department, puts it this way: "Secretary McNamara, in introducing P.P.B.S. in the Defense Department, had a definite managerial concept and strategy in mind. State is trying to formulate a managerial strategy at the same time as it is trying to develop a system to support that strategy." The question is, in the absence of a management strategy, will the computer—now an integral part of the P.P.B.S. system—be used merely to decorate and speed up already obsolete processes? Will information technology simply be applied to existing information-gathering processes? Will more information be gathered only to become useless because the persons who need it do not get it, or get it at the wrong time? P.P.B.S., which is principally concerned with planning and budgeting, is only a small part of this dilemma. The problem becomes more complex and urgent, say, in the implementation of policy or crisis management.

But those who plan carefully may be able to learn much from business experience with the application of advanced information technologies. There are some 35,000 computer systems operated by private industry in this country today. But in 1954—just thirteen years ago—only the first few were being installed for commercial use. The lessons have been and still are learned by business the hard way: mounting costs for useless data, duplication of functions and personnel, large-scale errors in business operations and decisions. The key problem resides in the inability or unwillingness of management to ask itself what it really wants from technology. What kind of information is needed by which persons at what times? What is the relation of the costs of this information to the benefits derived? More and more such questions are being formulated with insight and imagination and, as a result, the latest technological capabilities have made possible not only a change in the methods but also in the substance of business operations.

For example, certain items can now be mass-produced by inserting a magnetic tape into a computer which guides the machine in the manufacturing process. Instead of manufacturing such items at headquarters and shipping them where required, it may be cheaper to ship the magnetic tape and manufacture the items in the market areas. Imaginative thinking through of technology makes entirely new processes and procedures possible.

Banking systems, credit-card companies and airlines are among those operations whose present scope of service would be impossible without the relatively intelligent use of information technology. One can only imagine the chaos if, for some reason, governmental operations—from traffic control to internal revenue collection—were to return to the limitations imposed by precomputer administrative and clerical routines.

Today one can no longer think of just the computer. One must think in the more comprehensive terms of information technology or information systems. This fact is brought home dramatically by a review of costs. Ten years ago, the computer or central processor represented some 75 percent of the total cost of an automatic data-processing system. The so-called peripheral equipment—input/output devices, outside storage and communications links—accounted for some 25 percent. This is changing radically and by 1972 the cost relationships will be completely reversed. The cost of information processing and storage within the computer system will decrease 97 percent between 1963 and 1972, while the cost of communicating with the computer center will decrease by some 50 percent. The computer is emerging from its glassed-in throne room and, as it becomes increasingly accessible to those needing its services, the links between it and society proliferate both in number and in complexity.

This complexity itself simplifies the relationship between man and machine and makes the machine more and more an integral part of society. The information systems of today already are beginning to provide us with the ability to ask the computer questions through keyboard desk sets, light pencil drawings on a television-like screen, or still to a limited extent, by voice. Answers to such questions come back through a print or on a screen, or both. For instance, an engineer can make a sketch with a light pencil on the screen; the computer converts this sketch into a precise engineering draft which appears on the screen and can be rotated in perspective or altered at will. When the engineer gets what he wants he can either have the design printed out in hard copy or converted to a tape which runs a machine tool which, in turn, cuts the designed part out of metal. A typical multi-station system in a large corporation allows hundreds of managers across the country—and eventually across the world—to query a centrally located computer from their desk sets and receive instantaneous replies in visual form.

The key questions that have to be answered in order to build these systems and make them work usefully are: (1) Who needs the information? (2) What kind of information must be made available, in what detail, and how currently? (3) Must the system be complex enough to allow for machine guidance of the questioner if the question is unclear or unanswerable in the form presented?

In other words, what do we want from our technology? As our commercial systems are beginning to demonstrate, we can get what we want.

III

In the management of foreign affairs, information technology gives us usefully the chance to review what we are doing as well as what we want to do. I shall try to show that it will affect not only who makes the decisions or who is the instrument of response, as in the case of the Ambassador, but that for this and even more complex questions it will also change what the decisions are about. Further, it will determine whether decisions or conscious responses in particular instances are necessary at all, or are built into the system automatically.

The question of what we want raises, in turn, numerous questions which must be solved organizationally. Who will make the initial and continuing decisions on the data to be fed into the information system? How is data to be weighted for analysis and summary conclusions? Should more than one system be set up—for example, one for the State Department and one for the Central Intelligence Agency? Who should have access to the information from one or more systems? Should there be a switching center which controls who gets what?

However, I shall not concern myself here with these organizational questions. The answers to them will depend in large part on how we envisage the total impact of information technology on the substance of foreign-affairs management. The form we want the conduct of international relations to take—and we still have the weight in the world to shape that, if we assume the leadership—will have a profound effect on what the world looks like.

When Hitler embarked on the direct course leading to World War II, beginning with the announcement that Germany would rearm and culminating with the occupations of the Rhineland, Austria, the Sudetenland and Czechoslovakia, three principal arguments were made by those who counseled against intervention: (1) Hitler could not threaten Europe because Germany did not possess the means for all-out war and, therefore, he should be permitted to assert claims which might be legitimate; (1) Hitler already possessed enough power to make intervention too costly; and (3) Hitler, after he achieved Germany's immediate demands, would live in peace with his neighbors.

The first two arguments were based on information which was inadequate. The third argument was based on an inadequate appraisal of the man and of the psychological forces in Germany which supported him. The proper use of the kind of information and communications technologies now or soon to be available to us could have placed in perspective the first two arguments. Vast quantities of intelligence, most of it not secret but only undigested, on production, manpower, foreign trade, resources and technological probabilities could have provided the Allies at any stage with an accurate picture of German versus allied capabilities. The imponderables would have remained—questions about who would side with whom, about Hitler, the man, and the psychology of his nation—but even these could have been subjected to analysis aided by information technology. This is not to assert that history would necessarily have been changed; information can still be ignored or misused, and those who make policy are influenced by many factors, some of them essentially irrational. But technology cuts down the area of the unknown, narrows the basis for rational decision.

Many treaties are based on promises to do or not to do things which the partner cannot know about. This is so especially with nonaggression or disarmament treaties and their corollaries. If the ability to collect and process vast quantities of data, ranging from atmospheric samples to economic and transportation statistics, can give any one nation an increasingly accurate picture of trends and unusual activities in other nations, will the universal realization that others can divine a break in faith make such treaties obsolete? This could make the response of one nation to certain actions by others automatic, perhaps pre-programmed through simulation. Such "gaming" on the part of many competing powers could give them such an improved view of the possible consequences of their actions as to save them from many hazardous international experiments. Perhaps, in a crude way, this already is happening. The nuclear test-ban treaty might be considered just a formalized acknowledgement of mutually perceived facts. Can the use of information systems which are increasingly becoming more responsive and accurate push forward this kind of acknowledgement into broader areas of arms control and, someday, even make certain kinds of treaties obsolete?

Both in the negotiation of trade agreements and in their execution, an agreed-upon data base can make almost automatic the evaluation of the impact of concessions and of the responses to the withdrawal or tampering with concessions. Perhaps the principal function of the future trade negotiator will be, first, to arrive at an agreed-upon data base and, second, to negotiate on the basis of his evaluation of the national interest involved in facts known to all. On the other hand, it may be decided that, although this procedure would simplify one part of the negotiations, the facts are of such a proprietary nature as to preclude their use in this manner. In either event, the choice of action will be broadened significantly.

Undoubtedly, information systems for the conduct of foreign affairs will have to include major techniques for the forecasting of technological and socio-economic change. In order to prepare for the consequences of economic development in the emerging nations, and in the international exploitation of ocean, sub-arctic and extraterrestrial resources, substantial revisions in international law and economic policy obviously are going to be required. Information technology could be applied immediately to the collection of relevant socio-economic data both on the emerging nations and on newly developing resource areas, and eventually could relate them meaningfully to alternative courses of action—what kinds of investments should be made and by whom, what should be the distribution of costs and benefits, etc. On this basis of information, the substance of the decisions in these fields could be altered fundamentally. National and international concern could be concentrated on real issues and realistic alternatives.

These are but a few examples of how information technology may have an impact on the conduct of foreign affairs. The form and substance of what we do in this field are already changing. It is essential that we understand this and act upon the understanding systematically, imaginatively and with the best techniques available to us.

In the same decade that the new technology has emerged, the number of countries with which the United States conducts relations has more than doubled, the number of departments and agencies involved in foreign affairs has vastly increased and our sources of information have taken a quantum jump. The very process of decision-making has become infinitely complicated. Under these circumstances, the challenge of conducting our foreign affairs intelligently, of grounding policy on the best possible information, is a challenge to modern management and its use of organizational systems and technological tools. Is modern management now being applied to the conduct of United States foreign affairs? I think that a beginning has been made. Perhaps in this beginning we may also find that our statesmen—not the technicians, but those who must decide what is to be demanded of the technicians—have begun to think about what they want. For, once again, this is the central question: What do we want from our technology? If we know, we can get it.

The Privacy Question

CARL KAYSEN

Princeton University

Both the intellectual development of economics and its practical success have depended greatly on the large body of statistical information, covering the whole range of economic activity, that is publicly available in modern, democratic states. Much of this material is the by-product of regulatory, administrative, and revenue-raising activities of government, and its public availability reflects the democratic ethos. In the United States there is also a central core of demographic, economic, and social information that is collected, organized, and published by the Census Bureau in response to both governmental and public demands for information, rather than simply as the reflex of other governmental activities. Over time, and especially in the last three or four decades, there has been a continuing improvement in the coverage, consistency, and quality of these data. Such improvements have in great part resulted from the continuing efforts of social scientists and statisticians both within and outside the government. Without these improvements in the stock of basic quantitative information, our recent success in the application of sophisticated economic analyses to problems of public policy would have been impossible. Thus, the formation last year of a consulting committee composed largely of economists to report* to the Director of the Budget—himself an economist of distinction—on "Storage of and Access to Federal Statistical Data" was simply another natural

Reprinted from Carl Kaysen, "Data Banks and Dossiers," *The Public Interest,* 7 (Spring 1967), pp. 52–60, © by National Affairs, Inc. Used with the permission of the author and National Affairs, Inc.

* The full title of the Report, dated October, 1966, is: *Report of the Task Force on the Storage of and Access to Government Statistics,* and it is available from the Bureau of the Budget. The Committee which produced it were: Carl Kaysen, Chairman, Institute for Advanced Study; Charles C. Holt, University of Wisconsin; Richard Holton, University of California, Berkeley; George Kozmetsky, University of Texas; H. Russell Morrison, Standard Statistics Co.; Richard Ruggles, Yale University.

step in a continuing process. The participants were moved by professional concern for the quality and usability of the enormous body of government data to take on what they thought to be a necessary, important, and totally unglamorous task. They certainly did not expect it to be controversial.

The central problem to which the group addressed itself was the consequences of the trend toward increasing decentralization in the Federal statistical system at a time when the demand for more and more detailed quantitative information was growing rapidly. Currently, twenty-one agencies of government have significant statistical programs. The largest four of these—the Census, the Bureau of Labor Statistics, the Statistical Reporting Service, and the Economic Research Service of the Department of Agriculture—account for about 60 percent of a total Federal statistical budget of nearly $125 millions. A decade ago, the largest four agencies accounted for 71 percent of a much smaller budget. By 1970, the total statistical budget of the Federal Government will probably exceed $200 millions and, in the absence of deliberate countervailing effort, decentralization will have further increased. Yet, it has already been clear for some time that the Federal statistical system was too decentralized to function effectively and efficiently.

THE DRAMA BEGINS

Such is the background of the report which recommended the creation of a National Data Center. Here, Congressman Cornelius Gallagher (D., 13th District, N.J.) entered the scene, with a different set of concerns and objectives. He was Chairman of a Special Sub-committee on Invasion of Privacy, of the Government Operations Committee of the House, which held hearings on the proposed data center and related topics in the summer of 1966. To some extent the hearings themselves, and to a much greater extent their refraction in the press, pictured the proposed Data Center as at least a grave threat to personal privacy and at worst a precursor to a computer-managed totalitarian state. Congressman Gallagher himself saw the proposal as one more dreary instance of a group of technocrats ignoring human values in their pursuit of efficiency.

It now appears as if the public outcry which the Committee hearings stimulated and amplified has raised great difficulties in the way of the proposed National Data Center. To what extent are they genuine? To what extent are they unavoidable? Are they of such a magnitude as to outweigh the probable benefits of the Center?

In answering these questions, it appears simplest to begin with a further examination of the proposal itself. The inadequacies arising from our over-decentralized statistical system were recognized two decades ago; since then they have increased. The present system corresponds to an obsolete technology, under which publication was the only practical means of making information available for use. Publication, in turn, involved summarization, and what was published was almost always a summary of the more basic information available to the fact-gathering agency. In part, this reflected necessary and appropriate legal and customary restrictions on the Federal Government's publication of data on individuals or on single business enterprises. In part, it reflected the more fundamental fact that it was difficult or impossible to make use of a vast body of information unless it was presented in some summary form.

Any summarization or tabulation, however, loses some of the detail of the underlying data, and once a summary is published, retabulation of the original data becomes difficult and expensive. Because of the high degree of decentralization of the statistical system, it is frequently the case that information on related aspects of the same unit is collected by different agencies, tabulated and summarized on bases that are different and inconsistent, with a resultant loss of information originally available, and a serious degradation of the quality of analyses using the information. The split, on the one hand, between information on balance sheets and income statements, as collected by the Internal Revenue Service, and, on the other hand, the information on value of economic inputs and outputs as collected by the Census, is one example of this situation.

The result of all this is the substitution of worse for better information, less for more refined analysis, and the expenditure of much ingenuity and labor on the construction of rough estimates of magnitudes that could be precisely determined if all the information underlying summary tabulations were available for use. This, in turn, limits the precision of both the policy process, and our ability to understand, criticize and modify it.

These effects of the inability of the present system to use fully the micro-information fed into it are growing more and more important. The differentiation of the Federal policy process is increasing, and almost certainly will continue to do so. Simple policy measures whose effectiveness could be judged in terms of some overall aggregate or average response for the nation are increasingly giving way to more subtle ones, in which the effects on particular geographic areas, income groups, or social groups become of major interest. The present decentralized system is simply incapable of meeting these needs.

It is becoming increasingly difficult to make informed and intelligent policy decisions on such questions in the area of poverty as welfare payments, family allowances, and the like, simply because we lack sufficient "dis-aggregated" information—breakdowns by the many relevant social and economic variables—that is both wide in coverage and readily usable. The information the Government does have is scattered among a dozen agencies, collected on a variety of not necessarily consistent bases, and not really accessible to any single group of policy-makers or research analysts. A test of the proposition, for example, that poor performance in school and poor prospects of social mobility are directly related to family size would require data combining information on at least family size and composition, family income, regional location, city size, school performance, and post-school occupational history over a period of years in a way that is simply not now possible, even though the separate items of information were all fed into some part of the Federal statistical system at some time.

A secondary, but not unimportant gain from the creation of the data center, is in simple efficiency. At present, some of the individual data-collecting agencies operate at too small a scale to make full use of the resources of modern information-handling techniques. The use of a central storage and processing agency—while maintaining decentralized collection, analysis, and publication to whatever extent was desirable—would permit significant economies. As the Federal statistical budget climbs toward $200 million annually, this is not a trivial point. Even more important than prospective savings in money are prospective savings in the effort of information collection and the corresponding burdens on individuals, business, and other organizations in filling out forms and responding to questionnaires. As the demand for information grows, the need to minimize the costs in respondents' time and effort becomes more important. The present statistical system is only moderately well-adapted to this objective; a data center would make possible a much better performance on this score.

WHAT IT IS AND ISN'T

So much for the purpose of a data center; how would it function? First, it is important to point out that a data center is *not* the equivalent of single centralized statistical agency which takes over responsibility for the entire information-gathering, record-keeping, and analytical activity of the Federal government. Rather, it deals with only one of the three basic functions of the statistical system—integration and storage of information in accessible form—and leaves the other two—collection of information, and tabulation, analysis, and publication—in their present decentralized state. To be sure, if the Data Center is as effective and efficient as some of its proponents expect, some redistribution of the last set of tasks between the agencies presently doing them and the Center would probably occur. This, however, would be the result of choice on the part of the using agencies, if they saw an opportunity to do a better and less costly job through the Center than they could do for themselves.

The crucial questions, of course, are (a) what information would be put into the data center, and (b) how would access to it be controlled? In the words of the Task Force Report, the "Center would assemble in a single facility all large-scale systematic bodies of demographic, economic and social data generated by the present data-collection or administrative processes of the Federal Government, . . .integrate the data to the maximum feasible extent, and in such a way as to preserve as much as possible of the original information content of the whole body of records, and provide ready access to the information, within the laws governing disclosure, to all users in the Government, and where appropriate to qualified users outside the Government on suitably compensatory terms."

The phrase "large scale systematic bodies of demographic, economic and social data" translates, in more concrete terms, into the existing bodies of data collected by Census, the Bureau of Labor Statistics, the Department of Agriculture, the National Center for Health Statistics, the Office of Education, and so on. It also includes the large bodies of data generated as a by-product of the administration of the Federal income tax and the Social Security system. It does *not* include police dossiers from the FBI, personnel records of the Civil Service Commission or the individual government agencies, or personnel records of the armed services, and other dossier information, none of which fits what is meant by the phrase "large scale, systematic bodies of social, economic, and demographic data."

For the data center to achieve its intended purposes, the material in it must identify individual respondents in some way, by Social Security number for individuals, or an analogous code number now used within the Census for business enterprises called the Alpha number. Without such identification, the Center cannot meet its prime purpose of integrating the data collected by various agencies into a single consistent body. Whether these Social Security or Alpha numbers need in turn to be keyed to a list of respondents which identifies them by name and address within the data center itself, or whether that need be done only within the actual data collecting agencies, is a technical detail. That it must be done someplace is perfectly clear, as it now is done within the several agencies that collect the information.

On the other hand, it is not in general necessary that the central files in the data center contain a complete replica of every file on every respondent who has provided information to the original collectors. In many cases—e.g., the Social Security files—a properly designed sample would serve the same purposes more economically. To this extent, then, the data center will not contain a file on *every* individual, *every* household, *every* business, etc., but a mixture of a collection of samples—some of them relatively large—and complete files of some groups of reporting units which are particularly interesting and important from an analytical point of view. But here again, the significance of the difference between reproducing for the data center a complete file which already exists in some other agency, and reproducing only a sample therefrom, can easily be over-emphasized.

ANXIETIES

It is neither intemperate nor inappropriate to observe that the merits of the proposed data center have hardly been discussed in the tones that ordinarily mark consideration of a small change in government organization in behalf of greater effectiveness and efficiency. The anxieties stimulated by or crystallizing around the proposal can be divided into six groups: (1) the center will contain information that should not be in it; (2) the information can be improperly used by those within the government who have access to it; (3) the "bank" will be subject to cracking, so to speak, and data on individuals will be used to their detriment in any way from black-mail to gossip; (4) an enterprise of this sort is inherently expanding in nature, and no matter how modestly it begins, it will grow to include more and more, and eventually too much; (5) it both represents and encourages meddling and paternalistic government, trying to do too much in controlling the lives of its citizens; and (6) at a deeper level, it stands for a notion of an omniscient government, which is in some fundamental way inconsistent with our individualistic and democratic values. These categories are over-lapping in part and hardly all on the same logical level of discourse, but they seem to contain broadly all the criticisms that have been made.

To what extent are these problems real and new; to what extent are they simply translations into a new technical mode of familiar and persistent problems in the relation of citizens and government? And, if the latter, how well can variants of familiar mechanisms be adapted to deal with them? In what follows, I argue that while the fears raised by critics have real content, the problems are neither entirely novel, nor beyond the range of control by adaptations of present governmental mechanisms.

The first two questions go to the fundamental problem of government: *quis custodiet ipsos custodes?* The content of information now in the inven-tory of government agencies is controlled ultimately by the Congress, operat-ing through the appropriations process; and more immediately by the separate bureaucratic hierarchies of each data collecting agency, subject to the overall review of the Director of the Budget. He has a specific statutory responsibility for reviewing all governmental questionnaires directed to the public, with a view to eliminating duplication and keeping the total burden on respondents at a reasonable level. If this process seems to be working ineffectively, in the sense of ignoring persistent complaints, then the Appro-priations Subcommittees that deal with the budget requests of each data-collecting agency are readily able to exercise a further control. In practice, the existence of this restraint operates to reinforce powerfully the caution of the collecting agencies in expanding their requests.

A new data center would operate within the same framework of controls. Indeed, the Congress, in authorizing its creation, should define the kind of information which it would assemble, and could follow the line of demarca-tion of large-scale systematic demographic, economic, and social statistics suggested above. The inclusion of dossier information could be specifically prohibited. A clear distinction between "a dossier" and "statistical data file" on an individual can be made in principle; namely, for a dossier, the specific identity of the individual is central to its purpose; while for a file of data it is merely a technical convenience for assembling in the same file the connected set of characteristics which are the object of information. The purpose of the one is the assembly of information about specific people; the purpose of the other, the assembly of statistical frequency distributions of the many characteristics which groups of individuals (or households, business enterprises, or other reporting units) share. In practice, of course, this distinction is not self-applying, and administrators and bureaucrats, checked and overseen by politicians, have to apply it. But so is it ever.

A similar set of observations is relevant to the question of the control over the use of data in the center. The present law and practice governing the Census Bureau offer a model for this purpose. The law provides that information contained in an individual Census return may not be disclosed either to the general public or to other agencies of the government, nor may such information be used for law-enforcement, regulatory, or tax-collection activity in respect to any individual respondent. This statutory restriction has been effectively enforced, and the Census Bureau has maintained for years the confidence of respondents in its will and ability to protect the information they give to it. The same statutory restraint could and should be extended to the data center, and the same results could be expected of it. The data center would supply to all users, inside and outside the government, frequency distributions, summaries, analyses, but never data on individuals or other single reporting units. The technology of machine storage and processing would make it possible for these outputs to be tailored closely to the needs of individual users without great expense and without disclosure of individual data. This is just what is *not* possible under our present system.

TEMPTATIONS

However, it may be argued that the greater richness of the data file on any single reporting unit in a new data center as compared with those presently existing in the Census, the Social Security Administration, the Internal Revenue Service, and elsewhere would greatly increase the temptation for those with legitimate access to the data to use it improperly. The same argument goes to the third point listed above—the "cracking" of the center by outsiders ranging from corrupt politicians to greedy businessmen and organized criminals. It is clearly the case that centralized storage in machine-readable form of large bodies of information makes the rewards of successful abuse or "penetration" relatively large—compared to what they would be in a more decentralized, less mechanized system. It is not at all clear, however, that the cost of successful misapplication or penetration cannot be increased even more sharply than the rewards. In detail this is a technical problem of great complexity, but it seems clear from experience with a variety of secrecy-preserving techniques that a well-designed system of record storage and use could make "penetration" highly costly and to a large extent self-announcing. It is not difficult, for example, so to organize and code the basic records that programs for retrieving information routinely record the user and the purpose for which it was used. Any continued improper use would thus leave a trail that would invite discovery. Or, to mention another aspect, identifying numbers could be specially coded, and the key to that code made available on a much more restricted basis than were other codes. While no security system can be made perfect, it is feasible to make the costs of breaking it sufficiently high so as to keep the problem within tolerable bounds. The same kinds of safeguards would prevent misuse of the data by those with legitimate access to it.

The last three kinds of objections are similar in that they reflect a certain stance toward the government, and toward the evolution of its role in the larger society, and are not tied to any specific concrete problems. Indeed, the concrete problems underlying these broader concerns are those already examined. How will the contents of the data bank be controlled? Who will determine to what uses it may be put? How can we prevent the stock of information from being abused, misused, or simply misappropriated? But there appears beyond these specifics an attitude hard to discuss because of its intangible nature.

On the broadest level, one can simply reject the notion that there is an ineluctable ever-expanding process of governmental "intrusion" which must be resisted at every turn, yet inevitably overcomes whatever resistance the public offers. After all, this is the stuff of right-wing ideology. Opposed to this is the pragmatic liberal view that the public calls in the government, with more or less deliberation, when there are social problems to be solved which require governmentally-organized efforts and legally-enforceable obligations for their solution. Indeed, many proponents of this view see the restraints on government action built into our political system as too high, rather than too low, and action as typically too little and too late. On this view we have suffered more, at least in matters of domestic policy, from the feebleness of our government than from its overweening strength.

Without decisively choosing one over the other of these ideological stances, and with full recognition that a government too feeble for the welfare of its citizens in some matters may be too strong for their comfort or even their liberty in others, it is possible to believe, as I do, that the present balance of forces in our political machinery tends to the side of healthy restraint in matters such as these. After all, the very course of discussion on these problems, since the Center was first considered, supports this view. Accordingly, I conclude that the risky potentials which might be inherent in a data center are sufficiently unlikely to materialize so that they are outweighed, on balance, by the real improvement in understanding of our economic and social processes this enterprise would make possible, with all the concomitant gains in intelligent and effective public policy that such understanding could lead to.

The Privacy Question

DONALD N. MICHAEL

University of Michigan

In this paper, we are concerned with the future—the next twenty years or so. To look even that far ahead may well be a futile exercise, for the rate of change of technology and society threatens to make footless fantasying of any speculations about the impact of selected factors. However, those who will have significant influence on the political and social processes of the next twenty years are alive today. Consequently, they share to some large degree the values prevailing now, and this is important if we want to explore the significance of the computer for *our* kinds of privacy and freedom. The years beyond the realm of a twenty-year period, may find us dealing with a population a significant proportion of which holds values quite different from today's. Since we are not likely to know what those values might be, further speculations than we are about to undertake would hardly be worth the effort in the context of this Symposium.

One approach to the kind of speculations we shall pursue herein would be to review with great precision and perspicacity the history and ramifications of the concepts of privacy and personal freedom, and in this light, to look at the possible effects of computers on them. We will not follow this approach; space and the author's knowledge are too limited, and the concepts, whether they are refined philosophical, legal, ethical, or political formulations, undoubtedly will have their day when it is time to inhibit or facilitate the impact of the computer. Before then, the impact of computers on man will be reflected much more in the commonplace responses of our pluralistic society to these frequently misunderstood and misapplied concepts. In particular, it should be understood that the writer's grasp of these concepts is also of the "common" variety.

Reprinted from Donald N. Michael, "Speculations on the Relation of the Computer to Individual Freedom and the Right to Privacy," *The George Washington Law Review,* 33 (October 1964), pp. 270–286, by permission of the author and *The George Washington Law Review.*

This paper should be read, then, as no more than a stimulus to further speculation and much hard work. It is a preliminary exercise, an attempt to delineate some circumstances where computers and the concepts of personal freedom and privacy may come together in the day-to-day environment of the next couple of decades to enhance or detract from the practice and preservation of freedom and privacy. We shall concentrate on the role of computers as the technological agents for these developments. The microphone, tape recorder, miniature camera, and questionnaire are other formidable technological agents; their uses are well documented in two recent books and we need not review the matter here.[1] We shall be concerned more with the implications for privacy and freedom implicit in the means and capacities of computers for processing and evaluating information, however collected. Our goal is to identify the interactions and the circumstances to look for if we wish to anticipate the impact of computers on freedom and privacy.

In order to grasp fully the potential impact of computers, we must be clear about their versatility. In their simplest forms they can sort punched cards and perform, at high speed, routine arithmetical and statistical calculations. In their more elaborate versions, computers can

> be built to detect and correct errors in their own performance and to indicate to men which of their components are producing the error. They can make judgments on the basis of instructions programmed into them. They can remember and search their memories for appropriate data, which either has been programmed into them along with their instructions or has been acquired in the process of manipulating new data. Thus, they can learn on the basis of past experience with their environment. They can receive information in more codes and sensory modes than men can. They are beginning to perceive and to recognize.... Much successful work has been done on computers that can program themselves. For example, they are beginning to operate the way man appears to when he is exploring ways of solving a novel problem. That is, they apply and then modify, as appropriate, previous experiences with and methods of solution for what appear to be related problems. Some of the machines show originality and unpredictability.[2]

Let us also recognize that the impact of computer technology will not be unilateral. Rather, it will be profoundly affected by attitudes held by significant portions of the public and their leaders—attitudes favorable, indifferent, or antagonistic to privacy and freedom. There are, of course, great social pressures already operating which run counter to the preservation of privacy. We shall not explore the sources of these pressures and anxieties; they are recognized as chronic states of mind and action for a large part of our population and its leadership. But they result in conformity and in the justification of exposure, and in order to conform or to assure

---

1 Packard, The Naked Society (1963) and Brenton, The Privacy Invaders (1964).
2 Michael, Cybernation: The Silent Conquest 6–8 (1963).

that others meet certain standards of conformity, people need to know what other people are doing, especially in their less easily observable lives. Our mass media in particular stimulate and cater to this need, and they revel in the publicizing of personalities by stripping away privacy, whether it be from the individual, his home, the classified senatorial hearing, or the diplomatic conference. There is every reason to believe that defining "reality" in terms of persons and in personal terms will continue, especially as the new, depersonalized reality becomes too complex to convey much meaning to the average citizen. Personalities *are* meaningful, and defining reality in terms of personalities will continue both to appeal to the conventional wisdom and experience of most people and to provide an attitudinal environment wherein it is more permissible for business and government to probe persons, too.

Up to the present, "central city" concentration (in contrast to most suburban situations), population growth, and increasing physical mobility have given the individual some relative opportunity to lose himself, or to be anonymous, thereby preserving to some extent his privacy and his freedom of action. As we shall see, much of whatever ecological advantage these sources of anonymity provide probably will disappear even if, as is unlikely, the flight to the suburbs ceases.

We must realize, too, that the ways in which the applications of computer technology affect other important aspects of our social environment inevitably will reinforce or overcome attitudes about freedom and privacy. In particular, the computer will have an increasingly significant influence on the design and conduct of public policies. The states of mind and conditions for action resulting from the implementation of these policies will affect the ease with which one can pursue freedom and privacy. An obvious inhibiting influence upon that pursuit would be produced by the siege style of command and control of society which Harold Lasswell calls the "garrison state."[3] A garrison state might well be the consequence of an ever more elaborate proliferation of national security policies, guided and embellished by the kinds of computer-based war games, weapons systems, and sophisticated strategies which have become fashionable in the last several years. On the other hand, a federally integrated attack on crime, fully using the ability of the computer to organize and interpret data about criminals and crimes, eventually would free many terrorized people from threats of death or disaster and open business opportunities now preempted by the free-wheeling criminal. Thus, it would not be surprising if, in the future, people were willing to exchange some freedom and privacy in one area for other social gains or for personal conveniences. Nor would it be the first time they have done so.

With such background considerations in mind, let us speculate on particular circumstances in which the computer will confront what, in myth or actuality, we take to be present privileges of privacy or freedom.

[3] Lasswell, National Security and Individual Freedom 47–49 (1950).

PRIVACY

Consider that kind of privacy which exists by virtue of the ability to restrict access to information about oneself and one's related activities and records. By and large, the information thus restricted concerns the historical self: not only one's outward conduct, but also his inward evolution as a human being.

The availability of computers can alter seriously the degree to which one can restrict such access. Several factors which have determined degrees of privacy in the past are:

1) The ability of the privacy invader to bring together data which has been available, but which has been uncollected and uncollated;

2) The ability of the privacy invader to record new data with the precision and variety required to gain new or deeper insight into the private person;

3) The ability of the invader to keep track of a particular person in a large and highly mobile population;

4) The ability of the invader to get access to already filed data about the private person; and

5) The ability of the invader to detect and interpret potentially self-revealing private information within the data to which he has access.

What is the interplay of these factors and what is their significance for privacy in the light of the computer's capabilities? Much of one's privacy remains undisturbed because no one has had the ability to pull together available information—or because no one has been sufficiently interested to go to the trouble of doing so. To understand the private implications in available data might first require both locating and integrating much widely dispersed information.

The meaning of the information may be unclear, and, therefore, still private. More information may be needed, and the quality of it may depend on updated surveillance of the person involved. Considering the size and mobility of our society, these problems have made privacy invasion very difficult, but, as we shall see, the computer makes it much more feasible.

Private information about a person may exist which is ethically or legally restricted to those who have a legitimate right to it. Such information, about a great portion of our population, exists in business, medical, government, and political files, and in the form of psychological tests, private and government job application histories, federal and state income tax records, draft records, security and loyalty investigations, school records, bank records, credit histories, criminal records, and diaries. Each day more of these records are translated from paper to punchcards and magnetic tapes. In this way they are made more compact, accessible, sometimes more private, and, very importantly, more centralized, integrated, and articulated. The results are more complete records on each individual and a potential for more complete cross-correlations. The would-be invader who knows about these centralized or clustered inventories need not search for sources, and therefore he may be much more inclined to examine the records than if a major search for the sources of information were necessary.

As population and mobility increase, there will be other incentives to establish central data files, for these will make it easier for the consumer in new environments to establish who he is and, thereby, to acquire quickly those conveniences which follow from a reliable credit rating and an acceptable social character. At the same time, such central data files will make it easier for the entrepreneur or government official to ensure his security, since he will know at all times with whom he is dealing. In consequence, we can expect a great deal of information about the social, personal, and economic characteristics of individuals to be supplied voluntarily—often eagerly—in order that, wherever they are, they may have access to the benefits of the economy and the government.

While this sort of information is accumulating, the behavioral scientist, in direct consequence of the capabilities of the larger computer, will be improving his ability to understand, predict, and affect the behavior of individuals and groups. For the computer provides two prerequisites for the development of effective social engineering. First, only the computer can process fast enough the enormous amounts of data needed to know what the existing states of social and economic affairs are. In the past, such information was scanty or nonexistent, or what there was was more or less out of date. This no longer need be; and it certainly will not continue to be, since coping with sheer social complexity will require that such information be abundantly available. In the second place, the computer will let the social scientist manipulate enough variables and enough circumstances in sufficiently complex ways to invent subtle models about the behavior of man and his institutions. Simulation of the behavior of individuals and institutions through the use of computers is well under way, and all signs are that it will be exceedingly productive. There is every reason to believe, then, that with the development of these sophisticated models, and with access to centralized data banks where many of the characteristics of each person, the institutions with which he is involved, and the environment in which he operates are recorded, it should be possible to develop a sophisticated understanding of the present behavior of individuals and to predict with some assurance various aspects of their future behavior as well as to interpret and deduce aspects of their past behavior. How detailed and valid the conclusions will in fact be remains to be seen, but it is very likely that average citizens, as well as those who will have a vested interest in using such predictions, probably will overestimate their precision. Even today, many people are willing, indeed sometimes want, to believe that the behavioral scientist can understand and manipulate their behavior. Whether the ability to predict behavior will be used to invade privacy and freedom will depend on more than technological capability, but certainly the capacity for invasion will increase as behavioral engineering progresses.

The ability of the user of private information to gain access to already collected data about the individual will depend on several factors. To understand the range of possibilities, it should be recognized that, increasingly, data will be stored in memory banks shared by several users. A computer's memory banks can be so large that only a relatively few users can employ their entire capacity. Furthermore, the speed and capacity of computer processing is so high that it is much more efficient to have several users sharing time on the machine. This means that while one user's finger is moving toward a control button on his computer control panel, another person is using the circuits he will be using a few seconds later. In the future, one computer user might accidentally gain access, through equipment malfunction, to another's information stored in shared memory banks. It is also possible that such information could be retrieved deliberately and clandestinely by querying the computer with someone else's retrieval codes or by otherwise tricking the computer and its memories. Shared computer time and shared memory banks are new techniques, and the possibilities for their abuse will become fully apparent only with extensive use—and misuse. Of course, one information user, through legitimate means, could gain the use of information in another organization's memory. Certainly governmental interagency cooperation is inevitable. This is likely to occur also in some forms of interbusiness cooperation. As computers are interconnected more and more, as related organizations come to share specific memory banks, the inclination to share information and the ease of doing so will increase. After all, one of the advantages of the large scale use of computers is the savings to be made by eliminating duplication of information and by standardizing information collecting forms and data retrieval languages.

So far we have speculated on the increased opportunities which the computer provides for invading privacy from "outside" the person. But there are trends which suggest that many people are likely to *cooperate* in exposing their previously private selves. Systematic exposure of the private self through questionnaires, interviews, and test-taking is becoming steadily more widespread and probably more acceptable. The pressures toward—indeed, the attractiveness of—this kind of exposure are strong. In the first place, many Americans like to believe that getting ahead is a matter of ability and personality rather than luck or nepotism or some other kind of whimsey in those who hire and fire them, or in those who acknowledge or ignore them when opportunities arise. As the society grows more complex and the individual's sense of his ability to influence it in his own interest seems smaller, the tendency to depend for placement and advancement on what can be revealed about oneself which can be evidenced and acted on "scientifically" may well increase. This would be a natural extension of our dependence on the expert, for it will be the expert who will assess one's "true" value—at least one's economic value—by evaluating the private information one makes available about oneself. This response also will be a natural extension of our dependency on the machine, which in this case will help the expert or make the decisions itself about the value of the individual, impersonally but with great precision, on the basis of what it knows. And the machine will do so "privately"; it will not blabber secrets to other machines or other people.

Complementary pressures from those who would use information about the private person are likely to be great. The real or imagined need to use people efficiently will increase as more organizations find themselves in the throes of complicated and disrupting reorganization, remodeling people and procedures to meet requirements imposed by the use of automation and computers. Thus, executives and decision makers, responding to emotional and practical pressures, will try to squeeze the utmost from available personal information as clues to efficient job assignments. Increasingly, executives seem to seek security through technological intervention in the conduct of their activities. This tendency will be reinforced as more executives become the products of physical or social engineering backgrounds. As the behavioral scientists' predictive capabilities are increasingly recognized, and as business and government become more professionalized and rationalized through the combined impact of the availability and capabilities of computers and the hyperrational orientations of the personalities who tend these devices, management will seek more and more to learn all it can about the people it uses and about the people it serves, in order that its tasks may be more efficiently conducted in the public interest or in its own. This is, of course, simply a continuation of a well-rutted trend. For some years we have assumed or accepted that efficient government, government subject to as little internal disruption as possible, "requires" personnel selection on the basis of very private information about an applicant's sex life, family affairs, and early ideological enthusiasms. Since most new jobs in the past two decades have been government or government-related jobs, we can expect this trend to continue, the result being that more people than ever will be on file.

We have noted the likelihood that, as the years go on, the pressures will increase substantially to collect and use available data about the private individual, to enlarge the scope of that data, and to attempt to correlate and expand its implications and meaning. To some extent, each user of information will gather his own, but, in keeping with standards of efficiency, there will be efforts to gain access to data accumulated by others. For example, we can expect pressures to combine credit rating information with government job application information, with school psychological testing information, and so on. In an enormously complex society, everyone may have something to gain by this process as well as something to lose.

One thing is clear: for a long time indeed no correlation of data will account fully for the personality being evaluated and interpreted. Whatever the person providing the information believes, and whatever those using the information believe, there still will be a truly private person left, undetected by the computer. This is not meant in any mystical sense. It will take much longer than the twenty-year period we are dealing with to gain enough understanding of human beings truly to strip them of all the private self which they think they volunteer to expose. Nevertheless, in many situations, that information which can be processed through punchcards or memory tapes will be accepted as the important private profile of the individual. This will be so because, limited as the data may be in some abstract sense, this will be the information most conveniently available to the users for the assessment of the individual. And it will not be trivial or simple information. It will be impressive in its scope, and the computer will be impressive in the processing of it. Thus the users will choose to believe that this is *the* important part of the private life of the individual, and from the economic standpoint it may well be. Similarly, many of those supplying the information will come to believe they are revealing their private selves. In other words, that which will be valued and acted on as if it were *the* private individual will be that which can be tested and assessed in ways which can be recorded and manipulated by computers. Of course, not everyone will succumb to this bifurcation of self, but enough may do so to make it an important factor in our society. The result may be that many will feel they or others have no private lives. Others will feel that their "real" private lives are even more private because they are relatively more ignored—the computer won't be able to do anything with them. Thereby we shall have a new measure of privacy: that part of one's life which is defined as unimportant (or especially important) simply because the computers cannot deal with it.

No one using the output from a computer needs to know as much about the data fed into it as does the programmer. Without intimate and extensive understanding of the data and the uses to be made of it, the programs which determine how the computer operates, and hence the quality of its output, will be crude. On the other hand, executive decisions often depend less on knowledge of details than on overall grasp of the situation. As a result, the programmer often will be the person with potentially the most intimate knowledge of the private lives of those whose data is processed. This potentiality need not result in his having specific knowledge about specific people, since a programmer is unlikely ever to see the materials which are input to the computer whose processes he has arranged. But given his deeper understanding of how the data are being processed, what assumptions are made about the relationships among the data, what constraints must be put on the data in order for the computer to use it, it is entirely possible that the programmer may be called upon in difficult cases to enrich the executive's basis for decision making. In this way, the programmer may become privy to very private information about specific individuals. There may then arise a demand for programmers with ethical standards which now are not considered prerequisites to their trade. Inevitably, of course, there will be corruptibles among this group who will leak private information.

In another sense the programmer will become important for the preservation of privacy and freedom. The way he arranges the relationships in the information to be processed and the relative emphasis he gives to different items could result in distortions in the "history" of the person and, hence, in the implications of the data. In other words, the programmer could invent a private life. The question then arises of how the individual protects and asserts his own version of his private life over and against that defined by the computer. In the past, it has been possible to refer differences in present interpretations of past events to witnesses or paper records or photographs. Such records were public in that they were visually comprehensible. But records storage will become ever more bulky and retrieving information will become an increasingly awkward and vexing task as the population increases and as the amount of information about each member of the population increases for the reasons we have discussed. The incentives to put this information into computer memories will thereby increase. But if history is recorded on tapes, in magnetic codes, and on molecular films, the definition of what was will become ever more dependent on how the machine has been programmed and what it is able to retrieve from its memories. As respect for and dependence on the computer increases, it is likely that respect for and dependence on fragile and "ambiguous" paper records will decrease, lessening the ability of the individual to establish a past history different from that jointly provided by the programmer of raw data and the interpreter of processed data. There will be fewer opportunities to derive a public consensus on what the data "is," for there will be no public language in which the primary data will be recorded through which the public can verify the meanings and facts of the records. Robert Davis of the Systems Development Corporation compares this situation to the time when the Bible was interpreted to the illiterate, and what the Bible said and meant depended exclusively on what those who could read claimed it said and meant.[4]

On the other hand, centralization of private information and its preservation in computer memories may decrease illegitimate leaks of that information. Those who will have access to personal history will see much more of it than was usually the case when it was contained in printed records, but fewer curious eyes will have knowledge of any part of the private history of the individual.

PERSONAL FREEDOM

Now let us look at a few possible confrontations between the freedom of the individual and the computer.

There is one form of technology tied to the computer which today increases freedom for some and which may in the future decrease it for others. This is the technique of telemetering information from tiny sensors and transmitters embedded in the human body. Right now, one form of these devices keeps recalcitrant hearts beating steadily. In a few years, in variations of already existing experimental devices, they will transmit information about subtle internal states through a computer to the doctor, continually or at any time he wishes. Clearly, the lives and liberty of people dependent on such support will be enhanced, for it will provide greater opportunity to move and to live than would be theirs if this information were not so continuously and directly available.

It is not impossible to imagine that parolees will check in and be moni-

4 Interview With Robert Davis of Systems Development Corp., in Washington, D. C.

tored by transmitters embedded in their flesh, reporting their whereabouts in code and automatically as they pass receiving stations (perhaps like fireboxes) systematically deployed over the country as part of one computer-monitored network. Indeed, if they wish to be physically free, it is possible that whole classes of persons who represent some sort of potential threat to society or to themselves may be required to keep in touch in this way with the designated keepers of society.

It may seem farfetched to suggest that such people might walk the streets freely if their whereabouts and physiological states must be transmitted continually to a central computer. But two trends indicate that, at least for those who are emotionally disabled, this is not unlikely. We are now beginning to treat more and more criminals as sick people. We are beginning to commit them for psychiatric treatment rather than to jail. This treatment may have to continue indefinitely, since frequently a psychiatrist will not be prepared to certify that his patient will not commit the same kind of crime again (as is now required for sexual offenders under psychiatric treatment). At the same time, chemical and psychotherapeutic techniques for inducing tranquil emotional states are likely to improve. We may well reach the point where it will be permissible to allow some emotionally ill people the freedom of the streets, providing they are effectively "defused" through chemical agents. The task, then, for the computer-linked sensors would be to telemeter, not their emotional states, but simply the sufficiency of concentration of the chemical agent to ensure an acceptable emotional state. When the chemical agent weakens to a predetermined point, that information would be telemetered via the embedded sensors to the computer, and appropriate action could be taken. I am not prepared to speculate whether such a situation would increase or decrease the personal freedom of the emotionally ill person.

Already the computer is being used in conjunction with other technologies to retrieve information customarily stored in libraries. Doubtless, this use of the computer will expand greatly. Tied in with telephone lines and television cables, it will make it possible to gain access to vast areas of knowledge without leaving one's own local area. To the extent that knowledge increases the individual's opportunities for growth and effective mobility, we could say that such access-at-a-distance will increase his freedom. This will be especially true for those who do not live near the conventional repositories of information.

It is sometimes suggested that the computer will bring back town hall democracy by making it possible for every voter to express his opinion at the time his representative needs it, merely by a pushbutton response to a teletyped or mass media-transmitted request for the constituent's position on a given topic. The voter would gain more freedom to express himself, but that of the representative to act in terms of his own estimate of the best interests of the nation or his district might be lessened by such ubiquitous and massive grass roots expression. If the representative were able to determine the voter's understanding of the issue in order to assess the meaning of his pushbutton vote, however, and if he had information on voters so that they could be clustered according to background, thereby allowing patterns of votes to be more fully interpreted, the representative's ability to act in the combined interests of his various constituencies and the nation at large could be increased. Such data also would give him a better basis for providing his constituents with the information they need in order to vote more intelligently. The biggest unknown would seem to be whether one could count on developing an intelligent and enlightened public, or whether computers used this way would simply increase the likelihood of representatives' being swayed or dominated by a mass incapable of judging the meaning or implications of the complex issues it is asked to evaluate.

If it is worthwhile, the enormous capacity of computers can provide the basis for differentiating among many subpopulations. This capability could mean increased responsiveness on the part of data users and planners to the different social, psychological, and material needs of each of these populations. It could lead to more opportunities for individual expression, at least to the extent that the substance of individual expression is significantly differentiated by membership in various subgroups. In this sense, the computer could provide a greater opportunity for freedom than would be available in a large society which had to plan and operate in terms of overall averages rather than differentiated averages. But this capability for recognizing differences in populations and interest groups could also be applied to more detailed surveillance, causing a much greater loss of freedom than would result in a large population dealt with in overall averages. In the latter situation, the individual could more easily lose himself in the mass.

Behavior is internally mediated by the individual's history, personality, and physical capabilities, and externally by the constraints which environment imposes or the opportunities which it provides. Knowledge of external environmental characteristics increases the observer's ability to predict individual behavior. This is evidenced by the ability of police, military intelligence, or traffic control experts to predict, from knowledge of their environments, likely behavior of specified people or groups. Obviously, these predictions are improved greatly when added to knowledge of the environment is knowledge of the typical response of the individual to similar environments. The police capturing the criminal when he returns to the scene of the crime, the successful guerrila ambush, the parent removing "temptation" from the child's environment, and the "lead item" sale, are all examples of how the discovery of patterns of correspondence between environment and individual behavior toward it can be used to predict and channel that behavior. Better predictions about behavior should be possible when external environment can be codified and defined in much greater detail by using computerized data and by using the computer to detect patterns of sequences and arrangements of individual acts through elaborate analysis and synthesis of the data. The resulting predictions could be used to alter an environment (inanimate or human) in order to provide more opportunities for alternative behavior, to facilitate habitual behavior, or to inhibit or terminate behavior.

As usual, the consequences of environmental controls for freedom could go either way. Improved surveillance techniques would mean less crime or, at any rate, less than there would be without such techniques. Less crime means more freedom and privacy, at least for the law abider. But the same techniques could be used against the law abider if his government wanted to make, say, a routine security check in the interests of social stability.

When the application of computers requires that people change their behavior towards something familiar, they may well interpret this as an imposition on their freedom. This interpretation is in keeping with the belief long held by many that the machine is the chief threat to the spontaneity (freedom) of man. The recent furor over all-digit dialing demonstrates how seriously this threat is taken.[5] In the abstract, at least, one's freedom to dial long distance numbers direct may be increased by this new system, and certainly it is not lessened compared to what it was when one used a mixture of letters and numbers. But obviously many people feel their freedom has been abridged because for them it seems easier to remember combinations of letters and numbers, and because this change symbolizes more mechanization and, thereby, a challenge to the free man. Undoubtedly, there will be further "invasions" of this sort.

An important variant of this state of mind is found in responses to the nationwide computerized system which makes it possible for a cashier to determine quickly whether an unfamiliar person seeking to cash a check has a criminal record. Through this system (cashier-to-computer-to-police) a number of criminals have been apprehended while they waited for their check to be cashed. Abhorrence of the system and sympathy for the bum-check passer is a common—although, of course, not unanimous—response to descriptions of this system in action.

Apparently, in many minds there is combined a sense of "There but for the grace of God . . ." and a realization that the inclination to violence and lawbreaking (which most of us harbor) will be throttled more and more even in fantasy. For what is mere man against the implacable, all-seeing machine? The godlike omniscience of the computer essentially destroys his hope, and hence his freedom to fantasy, that he can get even unfairly with a society which he thinks has been unfair to him. If the computerized world of tomorrow produces the kinds of rationalized standards which increase one's frustration and inhibition, then certainly this invasion of one's right to hope (*i.e.*, to fantasy antisocial success) will be interpreted as some kind of invasion of his personal freedom. If so, there most certainly will be an acceleration of a trend already under way: "Frustrate" the machines. In a spirit of desperation and vengeance people are bending punchcards, filling prepunched holes, and punching out additional ones. (Injunctions have already made it clear that this destruction of private property will not be tolerated, regardless of its contribution to the preservation of psychic property; the machine wins.) They are also overpaying, by

5 See, e.g., *Time*, July 13, 1962, p. 53; *Washington Evening Star*, Jan. 31, 1964, p. A12, col. 2.

one cent, computer-calculated and computer-processed bills and refusing to use postal ZIP codes.

Now, it may well be that existing law or future decisions and actions of courts and legislatures will enforce and elaborate present legal powers in order to conquer the threats to freedom and privacy on which we have speculated. But seldom is a law promulgated in anticipation of problems, especially when there are powerful interests which benefit from freedom to exploit. Moreover, as we have seen, in most cases there may be a potential or actual gain for freedom or privacy along with the loss. And as we well know, even existing laws protecting privacy and freedom are often difficult to apply ubiquitously and effectively. In the hothouse world of Washington, D. C., it is commonly believed that anyone who is anyone at all has had or is having his phone tapped by government agents. Whether or not this is true, what is important is that people believe it is true, and they accept this situation feeling either that the government has a right to such spying or that, even if it hasn't, they can do nothing about it. We are all well aware of the increasing pressures to enlarge the search and arrest powers of the police in the face of expanding urban crime. We know, too, that in some places some of these powers have been granted or their unsanctioned use tolerated. And what shall we conclude about such modifications of the law as that represented by an executive order which gave the House Committee on Un-American Activities for the period of the 88th Congress the right of access to any income, excess-profits, estate, or gift tax records it wished concerning subversives?[6]

In the face of the increasingly complex tasks of maintaining, to say nothing of improving, urban society; in the face of general popular disagreement at best, and indifference at worst, about the proper conduct and protection of the individual in this alienated and splintered world; in the face of the special advantages the computer provides to increase information, command, and control—I would speculate that, at least for some time to come, these advantages will generally be considered more important than protection or preservation of those threatened aspects of freedom and privacy we have examined here.[7]

[6] Exec. Order 11109, 28, Fed. Reg. 5351 (1963).
[7] I am indebted to David T. Bazelon, Robert Davis, and Patricia McMonigle for their suggestions during the preparation of this article.

The Law

SALLY F. DENNIS

IBM Corporation

Whether or not we decide to put the law into the computer, it is now a fact that the computer is into the law. Such Federal Institutions as the Social Security Administration, Internal Revenue Service and Food and Drug Administration depend heavily on computers in order to perform their missions. Local government units use computers to maintain and search criminal records, traffic violations and motor vehicle registrations, as well as to manage land title records and courtroom bookkeeping and scheduling.

Law firms and lawyers soon may be using computers in their offices, although the computers may be physically elsewhere. The lawyer is likely to have a small device called a terminal actually in his office; it then will be connected via telephone lines to a remote computer, which the lawyer shares with other users.

One use of such a terminal may be the drafting of contracts, bills, amendments, patents, scholarly documents and the like. The role of the computer system is to store drafts of the documents, accept corrections and additions, and then to actuate the typewriter at the terminal to produce a clean, revised, justified and paginated draft. The same terminal can be linked to a business system when office accounting and invoicing are to be done, or into a courtroom scheduling system when it is useful to know the status of a pending case.

A few lawyers have proposed more exotic ways of using computers. Some of these are management of evidence in complex cases, analysis of the logic of contracts or laws, and prediction of judicial decisions. (The last suggestion caused heated controversy in the profession, apparently because some people interpreted it to mean that computers could make the decisions. This outsider has never understood why lawyers did not find it reassuring

Reprinted from Sally F. Dennis, "Shall We Put the Law Into the Computer?," *Law and Computer Technology*, 1 (January 1968), pp. 25ff, by permission of The World Peace Through Law Center.

to know that judges are sufficiently consistent in their thinking so that their decisions could be predicted.)

The Bar is asking questions about computers: Are computer records acceptable as legal evidence? Who bears liability for breakdowns and "bugs"? How do copyright laws apply to records that can be read by machines but not by eyes? Should computer programs be patented or copyrighted or either?

During the past decade, while lawyers and computers have been coming closer together in all of these ways, both lawyers and non-lawyers have proposed that we put the law itself into the computer and then use it to "find the law." Demonstrations have been performed and uncounted speeches, symposia, and papers—both scholarly and otherwise—have appeared. The notion of law in computers has been terribly IN. But, there are in motion actually only a few ventures in which the computer is used in some way to "find the law." Why, with all the talk, has there been no more action?

It appears that three interlinked problems currently block progress. The first is the state of the technology, the second is the economics of the undertaking, and the third is the nature of the individuals who may be better served by law in computers than by law in books.

To discuss technology, we need first to identify the possible reasons for putting law into computers. We could, for example, put it in and then simply print it out. The technology for doing this is readily available, but the process is hardly rewarding for law that already exists in print. For new law that will be printed by computer, it makes very good sense. We could store laws in machine-readable form so that later we could update and amend them; the technology presents no problems, and at least under some conditions the goal seems reasonable. We could use stored legal documents in a research environment to study linguistics—there is sufficient evidence on hand today to indicate that much can be learned about the structure of all language by studying such material with the aid of a computer.

TECHNOLOGY

Most of the people who talk about law in computers are thinking of something else—namely, legal research. If we store law with the expectation of using it for legal research, then we should have some reservations about technology as well as a clearer understanding of the behavior of search systems. There seems to be a mystique surrounding the whole idea of law in computers that polarizes at two extremes: One, the view that the machine's uncanny eye can in some unbelievable subtle fashion find anything and everything in its cache; and two, that once the precious words have vanished out of sight onto tape or disk, it is a gambler's game akin to fishing at the county fair to remake visible that portion of the store which is wanted.

Both views have some truth. The computer is just about infallible at recognizing information that is highly formatted, as, for example, Social

Security records. (Highly formatted means that certain information is present in every record, and that it is always present in the same place.) However, when the information is not highly formatted—as in statutes, cases, contracts, and the like—the infallible computer may be less than adequate because of natural variations in language.

To overcome language variations, relatively unformatted material has to be formatted in some way if it is to be selectively retrieved. This is done either by indexing in some fashion, or by storing the original full text and then constructing a directory which tells the location of each word in the text. If the material is pre-indexed, this is done by either a human or a computer program. Since indexing is as yet an inexact art, neither human nor computer program is yet infallible in this function. There is reason to hope that indexing may become more like an exact science as more effort is invested in learning how to do it automatically with machines, but there is still much room for improvement.

If the full text, no-index approach is used, when it comes time to retrieve a fallible human must decide for which words and phrases to search, and the computer simply scans for matching language. In effect, the problem of indexing has been merely transferred from indexer to searcher. Searching full text is a sound process if the human really wants to locate specific language, e.g., "res ipsa loquitur." However, if he wants to locate a concept, which may be not only somewhat vague but also phrased in many different ways, full text may be less useful.

In the IBM-American Bar Foundation joint project, in which we studied indexing and searching of the full text of 500 cases from the Northeastern Reporter, we learned that searching a machine-generated index derived from the text was more productive than searching the original text. Similar behavior has been noted by others working with files of scientific documents. The reason probably is that indexing serves to emphasize the important words in the text.

The mystique frequently asserts that full text in the computer is to be preferred over anything less, such as an abstract or index, because if it is all there the computer is sure to find it. This is an engaging oversimplification. Full text may very well be the best possible input to an automatic indexing system, but it is not the best possible input for searches for words and phrases—unless the words and phrases in themselves are what is desired from the search. Therefore, people who want to put law in computers should examine their goals. If the goal is to search legal literature for concepts, then when there exists a more or less exact (and hopefully automatic) science of indexing, the computer will be without question the place to store the law—or at least the index to the law.

Another reason for putting the law into the computer might be to compare the law of one governmental unit with that of another. If this is the purpose, we must retain the same reservations about technology as for legal research, because a method of relating comparable sections will have to be devised, and this inevitably will require some form of indexing.

If the prospective user of legal information can decide that he does not care about being certain of finding everything in the store that bears on a given concept, then it is realistic even with current technology to put the law into the computer. With systems currently workable the chances are high that he will find half or more of the applicable material, if he is willing to look through, say, 40 machine-selected documents. The important point is that he needs to recognize what degree of help he can expect from the computer.

Another question affecting both technology and economics is what law to include in the system. Some candidates might be the following:

Statutes; Appealed cases; Administrative rules and regulation; Agency decisions; Briefs; Municipal codes; International law; Bible, Talmud; Law encyclopedias; Treatises; Lower court cases; Law review articles; Private agreements; Research reports from law institutes.

ECONOMICS

Choosing the law to go into the system presupposes judgment as to the economic or other social value of the law. Which type of judgment should be made depends on how the system will be supported. If it functions as a commercial venture, then the economic value of the information to its users will have primary importance. If it is supported by some sort of subsidy, other values may be more important.

Along with choosing which law, it is necessary to choose how much law. The amount of law that goes into the system affects both technology and economics. It needs to be decided whether we absolutely have to go back all the way to the beginning, or whether a "law" of diminishing returns sets in. What, for example, is the relative value of pre-1900 law as compared with the law of this century?

THE HUMAN FACTOR

The third problem is the user. A well-known member of the Bar has recently inquired how, if law is put into computers, laymen can have access to it as they now have in books? That question needs to be resolved.

Even if we restrict our view of users to lawyers, we know that they vary markedly in what they think they want from a system. In the American Bar Foundation study, we tested computer responses to English prose legal questions by asking a panel of four lawyers to rate each response on the following scale:

A.—for a case sufficient alone to dispose of the problem;

B.—for a case insufficient by itself but which could be used with others to dispose of the problem;

C.—for a case in the subject area but which would not apply to the specific problem posed;

D.—for a case having no relation at all to the problem.

We found a surprising number of instances which for a given question and citation, the four lawyers on the panel managed to assign four different ratings. There were a number of times when three panel members would rate a case "D" and the fourth would rate it "A." Furthermore, there was no consistency as to which panel member was fourth.

To those of us who try to design effective systems, it seems to be evident that when we try to compare the performance qualities of different systems, we must test statistically. It also points up the likelihood that when and if computer-assisted legal search systems do become commonplace, there will be complainers. We can help to forestall the complaints by taking pains now to strip off the mystique whenever the opportunity presents itself.

These remarks do not lead to a loud, clear answer to the title question. There is no doubt that lawyers are already achieving positive gains through computer use; that as the usage of remote terminals grows, the gains will increase, and law in the computer will look better and better because the same terminals can also tie into the law. But right now, we have to say Yes, let's put the law into the computer, but with reservations and with more thought, planning, understanding and information than we have had in the past.

The Law

STEPHEN P. SIMS

Ferguson Personnel Agency, Inc.

Up to now the legal community has shown some but not a great deal of interest in the onslaught of computers. Several forums have been held but these have been directed in the main to the more mundane feature of the device, namely, its ability to facilitate searches by promoting a rapid and efficient information retrieval setup. This situation may indicate less of a lack of interest and more of a lack of knowledge on which to proceed.

Even the largest of the computer manufacturers is apparently without an appreciable file of actual occurrences and potentialities of computers in the law. Until several months ago, there were also apparently no court decisions having a computer or its influence as the central issue. Now there is one.

This is so even though the industry is growing at an alarming rate, will employ millions and have tens of thousands of units in use by firms of all kinds by 1970, affecting practically all our personal, business and professional needs.

This is so because, in spite of its size, the industry is new. What is needed now in the law is a forward-looking team to evaluate the capabilities of the computer and recommend the application of these capabilities to appropriate real problems in the law today.

This article is a comparatively minimal effort to attempt to select from six areas of the law current topics of interest which now or may shortly present opportunities for solution or a danger of intrusion by the computer. An attempt has been made to highlight but a single major issue within each topic which lends itself to automation. The use of the computer suggested within each topic is based on a machine capability as of this writing.

Reprinted from Stephen P. Sims, "Possible Legal Consequences of the Computer Explosion," *Data Processing Magazine,* 8 (July 1966), pp. 20–26, by permission of North American Publishing Co.

187

ANTI-TRUST

Fruitful avenues of trade practice inquiry may be opening to energetic government investigators because of the growing use of computers by major corporations.

The central computer installation, utilized by several competitors to reduce costs, might lead to a degree of cooperation in excess of legal bounds. It may be one matter to divulge sufficient information to a competitor in order to arrive at a useful processing program. It may be another that such a series of revelations might grow into a realization that cooperation in areas other than data processing may be preferable to open competition. Perhaps Adam Smith, if he could view some of the time-shared computer facilities of the future, would again say "People of the same trade seldom meet. . .but that the conversation ends in a conspiracy against the public. . . ."

The user who can afford to utilize fully his own computer and who some day goes beyond normal inventory and accounting applications may begin to program his machine to do certain useful things for his business which unfortunately are also illegal. For example, automatic refusals by the computer to process orders from preselected markets which coincidentally happen to be dominated by a competitor's product. Or perhaps an automatic recalculation of product prices based on a pre-selected competitor's scale and subsequent changes. Although this may sound bizarre, it may soon become standard practice for the government to zero in on programmers for pretrial interrogation. Instructions from superiors, although innocent-sounding to the programmer, may provide substantial evidence in an action against the corporation.

Assume a corporation does not share its computer with others and also avoids the obvious programming pitfalls as mentioned. Let us also assume this corporation is the largest in its relevant market. In the years to come, its state of automation is enhanced to the degree that it enjoys what may be termed a total management information system. The system, let us assume, was so expensive that none of its competitors could afford its equal. This corporation is thus placed in the enviable competitive position of being at least minutes "ahead" of its closest competitors, and in many important areas days "ahead."

Key executives at all points in the management ladder have machine-evaluated data placed at their fingertips in real time that enables them to make purchasing, scheduling, financing, employment, styling, processing, advertising, shipping and pricing decisions with unparalleled efficiency and speed. As a result of this competitive edge, this corporation shows growth greater than its competitors and in fact some fall by the wayside. The corporation continues to improve its information system, adding predictive programs developed by its huge technical staff based on the latest mathematical models of the national economy and consumer trends.

Let us assume that the sole difference between this firm and a competitor is its uniquely advanced computer system. The continued improvements give the firm a continued edge. Can it be said with certainty that this firm, assuming it thereafter attains monopoly proportions in its field, achieved its station through economic inevitability, or through the use of a modern marvel specifically designed to outdo the competition in every way possible? And if it is not yet a monopoly but its growth indicates it is so heading, is this firm in danger of being divested of this modern marvel because of the illegal path it lights before the corporate conveyance?

The growth of computer applicability is such that many firms, although too small to support their own installation and too disorganized or cautious to share one with their competitors, see its importance too and are demanding the advantages of automation. To service this class, the independent service bureau, not connected with any user or group of users, has been developed. To offer a wide range of computer services, a bureau must invest a great deal of time and dollars in the development of software, a library of computer programs covering the many types of input data and output informational needs of businessmen. It follows therefore that starting a substantial service bureau from ground zero, including, in addition to software, equipment, overhead and sales expenses, is quite an undertaking.

It also follows that an existing bureau finds expansion into new geographic markets difficult. Most are satisfied to remain neighborhood operations. The few that do expand usually find acquisition more feasible than internal growth through new offices. Since the service bureau industry is highly fragmented, possibly little danger of anti-trust violations exist in these many mergers. Let us assume, however, that one of the larger bureaus, firmly entrenched in many markets, decides to acquire, in one way or another, literally dozens of existing, highly successful bureaus across the land in areas new to the parent.

Let us assume that the new group can now effectively service any customer in any city in the United States, perhaps a feat no other independent bureau can match. Assume also that all acquisitions were made with an eye to keeping each noncompetitive with the next, in fact let us say that each acquired firm demanded such a division and worked with its competitors to assure nonoverlapping service markets. The question is, can this grouping (perhaps the most powerful) of existing bureaus run afoul of any existing legislation?

Whether performed under the label of merger or any other, can horizontal agreements to divide markets while sharing expertise and economies to collectively outdistance any competitor be permitted in light of existing legal precedent regarding the curtailment of tendencies to monopolize, or conspiracies which substantially restrain trade? Under any name, could a network of strong competitors, tied together for several years under mutually exclusive contracts designed to exploit a wide territory to the practical exclusion of any newcomer, fall under the prohibited "merger" category?

EVIDENCE

The expert witness is often a powerful force in the courtroom. His background, including academic and field achievements, qualify such an individual to provide the trier of fact with a useful opinion as to the outcome which should logically occur, given a certain set of facts at the outset. Consider the rather fantastic role the computer may, perhaps even today, play as an expert witness, called not by either party to the action, but perhaps by the court.

The computer can operate only on the facts it receives. Its program, once concluded, is a collection of logical operating instructions, based on recognized business or scientific formulae, as to its method of calculation on the facts it receives. The computer therefore cannot be influenced by outside factors nor can its input facts be colored, assuming proper supervision of preparation.

The court can thus have available an expert witness of its choosing in practically any area of human endeavor once the appropriate program based on accepted theory has been written prior to the time of trial. Although the later assumption may seem to be enormous, leaving the computer for the courts of the year 2000, consider the advances in programming now being made by dozens of manufacturers, service bureaus and consultants, in such diverse areas as automated media selection for large advertisers, optimized sales routing, consumer reaction modeling for new product research, optimum mix proportions for product blends based on current prices and quality, efficient construction scheduling of new plants, and more.

The systems designed for these complex tasks utilize in most instances long-accepted methods of calculation and prediction made useful only recently by the high speeds of the computer. Perhaps the same expertise employed by the analysts in building these programs could be useful in creating expert witnesses for the courts.

For example, most consumer product makers have detailed figures for their market densities over the nation. Their advertising programs are also carefully detailed. Their price changes are also centrally recorded. Their chances of obtaining certain levels of new market shares are also carefully calculated. Consider now a central expert witness computer brought in to testify in an antitrust case. Assuming all firms in the land could be organized to log their normally recorded business statistics such as those above in a standardized code form, the court could feed a defendant's statistics to its expert witness computer and, based on its generally accepted program, the computer would give forth concrete market share statistics and growth patterns of great value in predicting any overreaching by the firm in question. The defendant could of course feed the computer his version of the facts in rebuttal.

CONSTITUTIONAL LAW

The controversy developing between law enforcement and defense bodies over the necessity of proper counsel at the proper time will no doubt be resolved by our highest court in due course. Awaiting the arrival of practical tools for local enforcement of the court's decision will probably require much more patience. Perhaps the computer, with its capability of recording remote data, comparing and reporting on same, will provide our urban areas with a functioning system a lot sooner than expected.

Soon the timing of a confession with respect to time of arrest, entry into police headquarters and arraignment will be a mathematical certainty. Consider a network of time clock-punched card dispensing devices hung at each traffic light. When an arrest is made, the officer must pull a card from the nearest dispenser immediately. The time and location will be automatically punched on the card. The officer will then write in the arrest particulars; the accused will write in his personal particulars, at the same time noting the written warning on the card that whatever he says may be used against him.

Upon entry to the station house the arrest particulars will be punched into the same card and the card data fed to a remote memory unit. The final disposition of the prisoner, whether released or arraigned and held, will be fed in data form at the moment of the occurence to the same central memory. Thus, an accurate chronological history of the arrest, together with the timing of key details, such as call to attorney and appearance of attorney, will be available when necessary to establish the time lag inherent in many so-called coerced confessions, particularly in the absence of proper counsel. Citizen awareness of this type of system wherein the responsibility rests with the arresting officer to activate the time sequence promptly is of course its best guarantee for success.

Each of us value our vote highly. We consider its weight an absolute unit equal to the vote of any other citizen, even more so after the present wave of redistricting shall be completed. It is well known that electioneering within certain distances of each polling place is strictly prohibited. But each voter in the east has been violating this law in the last several national elections and coincidentally has enjoyed a vote with a weight greatly in excess of that of his western neighbor.

All television networks have constructed voting pattern models using selected districts that not only represent a broad cross-section of the population in every measurable way but also have remained basically unchanged through the last several elections. By plotting each district's voting habits against past election results a usually reliable prediction may be promulgated for a current election as soon as these particular districts close, their votes are tallied and compared with the historical data. This is usually done within minutes by computer.

What does a westerner who has not voted do when he hears his man is ahead or even declared a winner; does he stay home, confident of victory and does his neighbor then make an extra effort to get out and vote for the other side? Or do people jump on the band-wagon? Or what?

No one knows but could one disagree that, assuming westerners are not influenced both ways equally, the weight of the eastern vote is unfair? Is not the easterner electioneering within the very booth itself? Is this one man one vote or one man many votes? And all this occurs because the computer counts and compares in the same manner but faster than any politician ever did and the results are improperly flashed to polling areas that are not yet closed. The irony is that, by artificially influencing the western vote, the computer may be destroying its own model, thus rendering its own prediction useless.

TORTS

The numbers versus name game which caused excitement among the more independent citizens may cause one to question where the course of automation will ultimately carry our way of life and when it will become a violation of our right to privacy. Each of us now has a number designating our social security, driver's license, car registration, checking account, savings account, mortgage account, several credit cards, magazine subscription, telephone, utility, life insurance, health insurance, voter registration and dozens of others.

Plans are now under way which will provide for automatic debiting of one's pay to satisfy the family's basic creditors each week. The employer's account will merely suffer a debit also; the employee will never see a check. The individual will of course never carry cash since his entire life's purchasing and payment habits will be stored centrally for instant recall by any clerk in any store for instant credit.

The central computer will be programmed to keep a constant watch over your finances including the rate, type and location of your purchases. A danger signal will be emitted when necessary. This computer will be tied in to the state and federal governments, banks, utilities, insurance companies, hospitals and even remote free-standing input-output units on all highways. In this way, not only your purchases but tax habits, speeding tickets, health status, equity position, residence and incomes may be collected, collated, checked and cross-checked continuously in real time with all relevant agencies.

This will assure everyone that you are living your life reasonably and are in no danger of exercising that peculiar, disruptive and private right you had in the past, namely to have a private bank account which enabled you to have an occasional spree, a privately mortgaged home, a right to decide which creditor to pay when, and otherwise regulate your affairs with a minimum of outside assistance.

Complex computer systems which keep accurate, up to the second data on all citizens may seem repulsive to some and wonderful to others but consider this same system which errs. A computer report to a salesgirl in a department store is certainly a sight communication. The situation of a well-to-do gentleman being rejected before a group of friends by a snip of a salesgirl because of an incorrect report seems to be an unpleasant one at best.

Perhaps no actual damage would even have to be shown in order to collect a substantial judgement for libel. Similarly the improper return of one's checks to creditors, charging another's credit purchases to your account, listing an unheard of disease in your personal central computer file, or overlisting traffic violations resulting in a loss of driving privileges, would also appear to be grounds for a defamation action.

A recent highway demonstration of the versatility of computers resulted in a woman driver being presented before dozens of newspaper reporters as a law-breaker. Wide publicity was given to the demonstration. The driver was apparently looked upon by millions of readers as a criminal in flight even before she had an opportunity to plead and be tried. Again, a computer print-out, which could have been in error, signalled the arresting machinery into motion.

Where does computer product and service liability lie? Assuming that the central computer does err and that the error is clearly due to a mechanical failure. Is it with the bank, the computer manufacturer who built the machine and sold or leased it to the bank, or no one? Reason would dictate that the manufacturer of the chattel is liable for the foreseeable damage its machine may cause when produced in a careless manner.

Assuming the bank did not misuse the computer, perhaps the burden should lie with the manufacturer to show that reasonable care in production was employed even though this apparently unexplainable component failure occurred. Since the equipment is utilized in such sensitive areas perhaps absolute liability should be imposed with the manufacturer distributing its occasional losses through insurance with premiums added to price. If the personal injury of a libel may be in some way equated to the injury incurred by improperly produced food products, the idea is not too remote.

Next consider the consequences of a computer malfunction caused not by breakdown but by negligent human programming. Assuming the bank's own employee programmers commit the error, it seems clear that the bank is liable. However, assume the bank hired an outside programming consultant firm, an independent contractor, to write its library of programs. Who is liable? Was the bank free to delegate the programming responsibility? Should the bank have taken special precautions because of the injurious nature of the work? If the task is properly one for an independent contractor because of its special nature how can outside consulting firms protect themselves against such potentially awesome liabilities?

THE JUDICIARY

Judicial decisions are the foundation of living law and could one deny that individual judges draw upon their personal feelings many times in arriving at a decision when two paths are equally available. Before pressing an action, an attorney plots the historical judicial data relating to his case. If it is favorable, he continues. If not, he usually will not continue. At the conclusion of trial, a judge will draw on this same historical data which, together with his own background, will form the basis of his instructions or decision. Consider the application of computers to these areas.

Without too much difficulty, at least a general guideline may be plotted for both attorneys and judges in almost any area of case law. The factual surroundings, tenor of the times, extenuating circumstances, geographical location, judge, even jury make-up of any series of decisions may all be factors formulated as input to the computer.

Plotting this material as a trend would take years if done by hand. When using computers, the preparation of input data presents the only major area of time consumption. Sometimes the preparation of useful input is, however, impossibly complex. In most cases it is not and such computer use may even lead to predictions as to our future legal framework. It will no doubt settle many cases out of court. In some societies, jurimetrics may eventually lead to judgements by machine. In most it will be limited, as it should, to the role of a tool of the bar and court. There is no reason why it should not prove to be a highly useful one.

COMPETITION

Thousands of firms now use office copiers to reproduce material of interest to their employees from various copyrighted magazines, books and pamphlets. Within a limited area and number of copies, this use of such material would probably be considered fair. But consider the voluminous copying required to enable many information retrieval computer systems to operate efficiently. A corporation which now receives literature and distributes it through its library to its research staff on request could perform the same task more efficiently via computer.

By means of hard copy printouts or display readouts, one computer could provide the same article instantly to dozens of employees all across the country. Assuming the data printed or displayed had been obtained from copyrighted material, could this copying to place the material in memory and reproduction of it and distributing it to many remote locations be considered a violation of the copyright laws? To operate an effective computerized library containing all possible source materials, would the employer of the future be forced to obtain permission and rights from literally tens of thousands of authors?

If individuals outside the employing corporations begin to utilize the services of these advanced libraries, the conclusion seems more clear. Also consider the use of a copyrighted compilation of data which an independent computer service mathematically digests to form predictive guideposts and then sells its conclusions to users. Is this fair use? Or is this a reproduction of the original material in a revised form?

The large volume of trademarks registered with the patent office has prevented any systematic search prior to adopting one's own mark. Reliance on one's own knowledge, that of the patent office and one's competitor's has led to numerous lawsuits for alleged infringement that could have been avoided. The computer here presents an excellent method for prescreening in seconds all trademarks registered prior to the trademark one may have in mind. Advanced input devices to computers will accept an overlay composed of a halftone of the trademark in question separated into a fixed pattern of dots of varying gray. The entire overlay will be scanned and the resulting signals transmitted to a central computer.

The computer will then search its memory for a trademark of similar pattern quality. If one is found, it will be reprinted on a return screen immediately. Once the basic composition of a trademark is found to be similar to another, differentiating colors may then be checked out. The monumental task of photographing, screening and placing in memory all the presently registered trademarks will of course force this system to await implementation for some time.

It is difficult, however, for one to dream of the patent office operating in its present mode in the year 2000.

Other areas of the legal profession in which computer usage is now or may soon be involved follow; space limitations prevent more than their simple listing:

- Taxes; income, gift, death and other calculations made easy by computer.
- Estate planning and administration assistance; investment yield calculations and predictions.
- Labor-management relations; arbitration to reasonable compromise limits set by computer.
- Legal research; by holding, key word and other methods; automated libraries with microfilm.
- Information retrieval; systems for titles, patents and other documents.
- Teaching; improved methods based on computer studies of past student performances.

The flow of computers into the mainstream of legal thought is still several years away. The scarcity of actual court cases based on primarily a computer-based problem is widespread. Perhaps the first case to actually produce a decision regarding computers occurred several months ago in Nebraska. It held that computer magnetic tapes are acceptable as evidence under the business records exception to the hearsay rule. This article has been intended to ask many questions, initiate the flow as it were, while providing few, if any, concrete answers.

A discussion of computers in the law seems to reduce itself to a weighing of probabilities. In most cases the computer may be viewed as an arm of the governmental or corporate body, to be used for good or evil as conscience dictates. Apparently the computer and its power is being held in some awe by the courts, probably because most know so little about the subject.

Recently a programmer received a sentence of five years imprisonment for apparently turning over his employer's proprietary computer program to an outsider. The program must have been considered valuable indeed. As knowledge of the art of automation increases through the employment of responsible study and dissemination measures, the legal community will undoubtedly display less awe and greater eagerness toward these modern marvels, and we will all benefit.

V

THE CULTURE

Introduction

Because of its ability to calculate, to process information, and to simulate various conditions, the computer has become a powerful aid to man's analytic and reasoning capacities. In the sciences, the social sciences, and the humanistic disciplines, the computer is not only an important labor-saving device, but it also allows researchers to engage in some investigations that would not have been feasible by human labor alone. Moreover, it often produces results that are unexpected and lead to new discoveries; that is, it has a creative function as well. While some critics have claimed that the use of computers often debases research in the social sciences and the humanities by generating an exaggerated emphasis on the quantifiable, it is difficult to determine how much the fault lies with computers and how much it results from the desire of these disciplines to emulate the physical sciences. As Jacques Barzun, himself a staunch critic of the use of computers in such research, has pointed out: "What have the humanists been doing for thirty-five years except to do exactly what a computer would do, only with their own unaided card indexes and fountain pens? They have taken apart poetry, they have taken apart novels, they have counted images, they have followed symbols that are sometimes non-existent, they have destroyed their own subject matter by a pseudo-computer-like approach, and now they have only themselves to blame if they have to learn the tricks and the jargon of computerizing."[1]

It should not be denied, however, that tools help to shape the nature of what men do. Just as in the economic and political spheres new types of activities have been engendered because of the existence of computer technology, likewise in scholarly research, the kinds of questions that are posed are often shaped, at least in part, by the desire to exploit the new technology. Thus, researchers may construct hypotheses which include many more variables than they would have considered in the days before the computer; and they will entertain calculations and statistical manipulations which would not have been deemed important previously.

The impact of the computer upon our culture derives not only from its uses in research, but also from its influence on our modes of thinking, our

[1] Jacques Barzun, "Computers for the Humanities? A Record of the Conference Sponsored by Yale University on a Grant from IBM, January 22–23, 1965" (New Haven, Connecticut: Yale University, 1965), p. 149.

199

vocabulary and symbolism. Such notions as "feedback" and "input and output" have become fairly widespread and exert some influence on the way in which men think about making decisions. The man-like powers of the computer and its potentials in education and in disseminating information to the population at large are also important cultural effects. While its potential as a tool in the educational process has not yet been fully exploited, its ability to engage in operations that approximate "human reasoning" has generated much popular agitation—ranging from awe and admiration to fear and resentment. Both of these extreme reactions will, no doubt, become muted as an increasingly greater proportion of the population comes into closer contact with the operation of computers.

Some analysts have predicted that in the future large numbers of people may be able to consult the computer much as they now consult dictionaries or encyclopedias. "A combination of the telephone, the computer, and a variety of print, visual, and audio outputs" may eventually render "the mass media archiac,"[2] as individuals ask for and retrieve news stories, dramas, or other types of information and entertainment which interest them. Such individualization is also the aim of those who are attempting to introduce computers into the schools. However, here as in other areas of computer application, the eventual outcome will hinge upon the pre-existing social structures and attitudes and the capacity or willingness to make the changes that would be requisite for such individualization to come about.

The selection by Patrick Suppes discusses this promise of individualized instruction. The technology, Suppes argues, is already available, although preparing curriculum for individualized instruction is expensive and as yet there has been little "operational experience" in how to do it. Nevertheless, he predicts that "within the next decade many children will use individualized drill-and-practice systems in elementary school; and by the time they reach high school, tutorial systems will be available on a broad basis."

Anthony G. Oettinger emphatically disagrees. He maintains that "education's institutional rigidity combined with infant technology's erratic behavior preclude really significant progress within the next decade, if significant progress is interpreted as widespread and *meaningful* adoption, integration, and use of technological devices...within the schools."

The essay by George A. Miller points out that insofar as machines and organisms perform the same function, they are "instances of theoretical systems of far greater generality." The cyberneticist attempts to discover the general principles governing these larger systems. Today, Miller notes, "we are...about as far from building a computer modelled on the human brain as Archimedes was from building an atom bomb...we do not yet know enough about the brain or the principles on which it operates." But by analyzing how man's brain works and what functions man can best perform, "we will gain increasingly deeper insight into how best to use computers to perform functions that are difficult for him."

[2] Ithiel de Sola Pool, "Social Trends," *Science & Technology,* 76 (April 1968), p. 94.

R. W. Hamming discusses the ways in which the use of computers can help to sharpen men's thinking. He comments that "I have...on several occasions examined a problem as if I were going to put it on a computer, though I had no intention of actually doing so, and in the examination found the answer I was looking for."

Computers extend man's thinking in the sciences by acting both as "instrument" and as creative "actor," Oettinger explains. To illustrate this dual function of the computer in scientific research, he explores the use of the computer as an instrument in determining the structure of protein molecules and as an actor in the construction of functional models.

Turning to the social sciences, Ithiel de Sola Pool examines the use of computers in the simulation of total societies. A simulation attempts to make a conditional prediction, looking at what would happen under some hypothetical future circumstances. In such simulations, the computer forces the researcher to think more precisely. "Typically," Pool notes, "if a social scientist says there is a tendency for Y to go one way if X goes a certain way, he makes it clear that he is only singling out one of its determinants in a statement which is true only insofar as other things remain constant. The simulator, on the other hand, attempts to bring in all the variables that he considers to be important in determining the outcome, to state the relationships among them with such completeness that only one degree of freedom remains, and in that way to state what the actual outcome would be if a particular variable X were changed in a certain way." Often, however, either the data or the theory is lacking for such simulations.

The uses of the computer in humanistic research are similar to those in the sciences and the social sciences. The selection by John R. Pierce, however, examines another aspect of computer applicability in the humanities: the use of the computer to create art. "In principle," says Pierce, "what makes Mozart like Mozart, Haydn like Haydn, Wagner like Wagner is not beyond analysis. I would be very surprised if someone could cause a computer to produce good and original Mozart at the push of a button. I wouldn't be surprised at someone's making the computer sound something like Haydn or Mozart or Bach." The computer enables the artist "to create within any set of rules and any discipline he cares to communicate to the computer. Or, if he abandons discipline, he may leave everything to chance and produce highly artistic noise."

J. C. R. Licklider takes a futuristic look at computerized information systems that might replace conventional libraries. "In the present century," he argues, "we may be technically capable of processing the entire body of knowledge in almost any way we can describe." But there are problems associated with the high cost of such operations, the large memories required, and the difficulties of developing an "unambiguous English."

If the potential uses of the computer appear to be almost limitless, what kinds of attitudes prevail among the population about this incredible instrument? Robert S. Lee reports on a study of 3,000 interviews with a cross-section of the American public concerning attitudes towards the computer. The "appreciation of the computer as a beneficial tool" is found to be "highest among people who are familiar with the world of business, among those who have an interest in mechanical things, among people who have an optimistic outlook, and among those who are generally openminded and receptive to things that are new and different." But there is also fear stemming from the notion that "the computer is an autonomous thinking machine." The data show that familiarity reduces the sense of uneasiness. Therefore, Lee concludes, "it will probably be the youngsters who grow up in a world of computers who will more truly understand the new relationship of man and the machine."

The relationship of man and the machine is the subject of Bruce Mazlish's essay. Man's refusal to acknowledge that there is no discontinuity between man and machines, Mazlish contends, lies at the base of the "distrust of technology." Man is "continuous with the tools and machines he constructs." This is a difficult concept to accept. Nevertheless, the man-machine discontinuity is likely to be abandoned, just as the major discontinuities of an earlier day have been discarded: between man in relation to the universe (broken by Copernicus), between man and the animal kingdom (broken by Darwin), and between man and his more primitive state (broken by Freud).

Marcia Ascher's exploration of the science fiction literature pertaining to computers reveals the following major themes: giving instructions to computers requires great care (because they are taken literally); creativity always remains in man; and the responsibility for decision-making must remain clearly in the hands of men. The stories contain the suggestion that "impersonal treatment and almost impossible communication are associated with companies or agencies that use computers." Ascher points out that there is a real danger that the speed of the computer may lead to decision-making that is based on insufficient contemplation.

FOR FURTHER READING

Edmund A. Bowles, ed., *Computers in Humanistic Research* (Englewood Cliffs, N.J.: Prentice-Hall, Inc., 1967).

Ithiel de Sola Pool, "Social Trends," *Science & Technology,* 76 (April 1968), pp. 87–101.

E.A. Feigenbaum and Julian Feldman, eds., *Computers and Thought* (New York: McGraw-Hill Book Co., Inc., 1963).

Donald N. Michael, *The Unprepared Society: Planning For A Precarious Future* (New York: Basic Books, Inc., 1968).

Anthony G. Oettinger, *Run, Computer, Run: The Mythology of Educational Innovation* (Cambridge, Mass: Harvard University Press, 1969).

Harold Sackman, *Computers, System Science, and Evolving Society* (New York: John Wiley & Sons, 1967).

PATRICK SUPPES

Stanford University

Current applications of computers and related information-processing techniques run the gamut in our society from the automatic control of factories to the scrutiny of tax returns. I have not seen any recent data, but we are certainly reaching the point at which a high percentage of regular employees in this country are paid by computerized payroll systems. As another example, every kind of complex experiment is beginning to be subject to computer assistance either in terms of the actual experimentation or in terms of extensive computations integral to the analysis of the experiment. These applications range from bubble-chamber data on elementary particles to the crystallography of protein molecules.

As yet, the use of computer technology in administration and management on the one hand, and scientific and engineering applications on the other, far exceed direct applications in education. However, if potentials are properly realized, the character and nature of education during the course of our lifetimes will be radically changed. Perhaps the most important aspect of computerized instructional devices is that the kind of individualized instruction once possible only for a few members of the aristocracy can be made available to all students at all levels of abilities.

Because some may not be familiar with how computers can be used to provide indidualized instruction, let me briefly review the mode of operation. In the first place, because of its great speed of operation, a computer can handle simultaneously a large number of students—for instance, 200 or more, and each of the 200 can be at a different point in the curriculum. In the simplest mode of operation, the terminal device at which the student sits is something like an electric typewriter. Messages can be typed out by the

Reprinted from Patrick Suppes, "Computer Technology and the Future of Education," *Phi Delta Kappan,* 42 (April 1968), pp. 420–423, by permission of Phi Delta Kappa, Inc.

computer and the student in turn can enter his responses on the keyboard. The first and most important feature to add is the delivery of audio messages under computer control to the student. Not only children, but students of all ages learn by ear as much as by eye, and for tutorial ventures in individualized instruction it is essential that the computer system be able to talk to the student.

A simple example may make this idea more concrete. Practically no one learns mathematics simply by reading a book, except at a relatively advanced level. Hearing lectures and listening to someone else's talk seem to be almost psychologically essential to learning complex subjects, at least as far as ordinary learners are concerned. In addition to the typewriter and the earphones for audio messages, the next desirable feature is that graphical and pictorial displays be available under computer control. Such displays can be provided in a variety of formats. The simplest mode is to have color slides that may be selected by computer control. More flexible, and therefore more desirable, devices are cathode-ray tubes that look very much like television sets. The beauty of cathode-ray tubes is that a graphical display may be shown to the student and his own response, entered on a keyboard, can be made an integral part of the display itself.

This is not the place to review these matters in detail; but I mean to convey a visual image of a student sitting at a variety of terminal gear—as it is called in the computer world. These terminals are used to provide the student with individualized instruction. He receives information from audio messages, from typewritten messages, and also from visual displays ranging from graphics to complex photographs. In turn, he may respond to the system and give his own answers by using the keyboard on the typewriter. Other devices for student response are also available, but I shall not go into them now.

So, with such devices available, individualized instruction in a wide variety of subject matters may be offered students of all ages. The technology is already available, although it will continue to be improved. There are two main factors standing in our way. One is that currently it is expensive to prepare an individualized curriculum. The second factor, and even more important, is that as yet we have little operational experience in precisely how this should best be done. For some time to come, individualized instruction will have to depend on a basis of practical judgment and pedagogical intuition of the sort now used in constructing textbook materials for ordinary courses. One of the exciting potentialities of computer-assisted instruction is that for the first time we shall be able to get hard data to use as a basis for a more serious scientific investigation and evaluation of any given instructional program.

To give a more concrete sense of the possibilities of indidualized instruction, I would like to describe briefly three possible levels of interaction between the student and computer program. Following current usage, I shall refer to each of the instructional programs as a particular system of instruction. At the simplest level there are *individualized drill-and-practice systems,* which are meant to supplement the regular curriculum taught by the teacher. The introduction of concepts and new ideas is handled in conventional fashion by the teacher. The role of the computer is to provide regular review and practice on basic concepts and skills. In the case of elementary mathematics, for example, each student would receive daily a certain number of exercises, which would be automatically presented, evaluated, and scored by the computer program without any effort by the classroom teacher. Moreover, these exercises can be presented on an individualized basis, with the brighter students receiving exercises that are harder than the average, and the slower students receiving easier problems.

One important aspect of this kind of individualization should be emphasized. In using a computer in this fashion, it is not necessary to decide at the beginning of the school year in which track a student should be placed; for example, a student need not be classified as a slow student for the entire year. Individualized drill-and-practice work is suitable to all the elementary subjects which occupy a good part of the curriculum. Elementary mathematics, elementary science, and the beginning work in foreign language are typical parts of the curriculum which benefit from standardized and regularly presented drill-and-practice exercises. A large computer with 200 terminals can handle as many as 6,000 students on a daily basis in this instructional mode. In all likelihood, it will soon be feasible to increase these numbers to a thousand terminals and 30,000 students. Operational details of our 1965-66 drill-and-practice program at Stanford are to be found in the book by Suppes, Jerman, and Brian.[1]

At the second and deeper level of interaction between student and computer program there are *tutorial systems,* which take over the main responsibility both for presenting a concept and for developing skill in its use. The intention is to approximate the interaction a patient tutor would have with an individual student. An important aspect of the tutorial programs in reading and elementary mathematics with which we have been concerned at Stanford in the past three years is that every effort is made to avoid an initial experience of failure on the part of the slower children. On the other hand, the program has enough flexibility to avoid boring the brighter children with endlessly repetitive exercises. As soon as the student manifests a clear understanding of a concept on the basis of his handling of a number of exercises, he is moved on to a new concept and new exercises. (A detailed evaluation of the Stanford reading program, which is under the direction of Professor Richard C. Atkinson, may be found in the report by Wilson and Atkinson.[2] A report on the tutorial mathematics program will soon be available. The data show that the computer-based curriculum was particularly beneficial for the slower students.)

At the third and deepest level of interaction there are *dialogue systems*

[1] P. Suppes, M. Jennan, and D. Brian, *Computer-assisted Instruction at Stanford: The 1965–66 Arithmetic Drill-and-Practice Program,* New York: Academic Press, 1968.
[2] H. A. Wilson and R. C. Atkinson, *Computer-based Instruction in Initial Reading: A Progress Report on the Stanford Project.* Technical Report No. 119, August 25, 1967, Institute for Mathematical Studies in the Social Sciences, Stanford University.

aimed at permitting the student to conduct a genuine dialogue with the computer. The dialogue systems at the present time exist primarily at the conceptual rather than the operational level, and I do want to emphasize that in the case of dialogue systems a number of difficult technical problems must first be solved. One problem is that of recognizing spoken speech. Especially in the case of young children, we would like the child to be able simply to ask the computer program a question. To permit this interaction, we must be able to recognize the spoken speech of the child and also to recognize the meaning of the question he is asking. The problem of recognizing meaning is at least as difficult as that of recognizing the spoken speech. It will be some time before we will be able to do either one of these things with any efficiency and economy.

I would predict that within the next decade many children will use individualized drill-and-practice systems in elementary school; and by the time they reach high school, tutorial systems will be available on a broad basis. Their children may use dialogue systems throughout their school experience.

If these predictions are even approximately correct, they have far-reaching implications for education and society. As has been pointed out repeatedly by many people in many different ways, the role of education in our society is not simply the transmission of knowledge but also the transmission of culture, including the entire range of individual, political, and social values. Some recent studies—for example, the Coleman report—have attempted to show that the schools are not as effective in transmitting this culture as we might hope; but still there is little doubt that the schools play a major role, and the directions they take have serious implications for the character of our society in the future. Now I hope it is evident from the very brief descriptions I have given that the widespread use of computer technology in education has an enormous potential for improving the quality of education, because the possibility of individualizing instruction at ever deeper levels of interaction can be realized in an economically feasible fashion. I take it that this potentiality is evident enough, and I would like to examine some of the problems it raises, problems now beginning to be widely discussed.

Three rather closely related issues are particularly prominent in this discussion. The first centers around the claim that the deep use of technology, especially computer technology, will impose a rigid regime of impersonalized teaching. In considering such a claim, it is important to say at once that indeed this is a possibility. Computer technology could be used this way, and in some instances it probably will. This is no different from saying that there are many kinds of teaching, some good and some bad. The important point to insist upon, however, is that it is certainly not a *necessary* aspect of the use of the technology. In fact, contrary to the expectations sometimes expressed in the popular press, I would claim that one of the computer's most important potentials is in making learning and teaching more personalized, rather than less so. Students will be subject to less regimentation and lockstepping, because computer systems will be able to offer highly individualized instruction.

The routine that occupies a good part of the teacher's day can be taken over by the computer.

It is worth noting in this connection that the amount of paper work required of teachers is very much on the increase. The computer seems to offer the only possibility of decreasing the time spent in administrative routine by ordinary teachers. Let us examine briefly one or two aspects of instruction ranging from the elementary school to the college. At the elementary level, no one anticipates that students will spend most of their time at computer consoles. Only 20 to 30 percent of the student's time would be spent in this fashion. Teachers would be able to work with classes reduced in size. Also, they could work more intensely with individual students, because some of the students will be at the console and, more importantly, because routine aspects of teaching will be handled by the computer system.

At the college level, the situation is somewhat different. At most colleges and universities, students do not now receive a great deal of individual attention from instructors. I think we can all recognize that the degree of personal attention is certainly not less in a computer program designed to accommodate itself to the individual student's progress than in the lecture course that has more than 200 students in daily attendance. (In our tutorial Russian program at Stanford, under the direction of Joseph Van Campen, all regular classroom instruction has been eliminated. Students receive 50 minutes daily of individualized instruction at a computer terminal consisting of a teletype with Cyrillic keyboard and earphones; the audio tapes are controlled by the computer.)

A second common claim is that the widespread use of computer technology will lead to excessive standardization of education. Again, it is important to admit at once that this is indeed a possibility. The sterility of standardization and what it implies for teaching used to be illustrated by a story about the French educational system. It was claimed that the French minister of education could look at his watch any time of the school day and say at once what subject was being taught at each grade level throughout the country. The claim was not true, but such a situation could be brought about in the organization of computer-based instruction. It would technically be possible for a state department of education, for example, to require every fifth-grader at 11:03 in the morning to be subtracting one-fifth from three-tenths, or for every senior in high school to be reciting the virtues of a democratic society. The danger of the technology is that edicts can be enforced as well as issued, and many persons are rightly concerned at the spectre of the rigid standardization that could be imposed.

On the other hand, there is another meaning of standardization that holds great potential. This is the imposition of educational standards on schools and colleges throughout the land. Let me give one example of what I mean. A couple of years ago I consulted with one of the large city school systems in this country in connection with its mathematics program. The curriculum outline of the mathematics running from kindergaten to high school was excellent. The curriculum as specified in the outline was about as good as any in the country. The real source of difficulty was the magnitude of the discrepancy between the actual performance of the students and the specified curriculum. At almost every grade level, students were performing far below the standard set in the curriculum guide. I do not mean to suggest that computer technology will, in one fell stroke, provide a solution to the difficult and complicated problems of raising the educational standards that now obtain among the poor and culturally deprived. I do say that the technology will provide us with unparalleled insight into the actual performance of students.

Yet I do not mean to suggest that this problem of standardization is not serious. It is, and it will take much wisdom to avoid its grosser aspects. But the point I would like to emphasize is that the wide use of computers permits the introduction of an almost unlimited diversity of curriculum and teaching. The very opposite of standardization *can* be achieved. I think we would all agree that the ever-increasing use of books from the sixteenth century to the present has deepened the varieties of educational and intellectual experience generally available. There is every reason to believe that the appropriate development of instructional programs for computer systems will increase rather than decrease this variety of intellectual experience. The potential is there.

The real problem is that as yet we do not understand very well how to take advantage of this potential. If we examine the teaching of any subject in the curriculum, ranging from elementary mathematics to ancient history, what is striking is the great similarity between teachers and between textbooks dealing with the same subject, not the vast differences between them. It can even be argued that it is a subtle philosophical question of social policy to determine the extent to which we want to emphasize diversity in our teaching of standard subjects. Do we want a "cool" presentation of American history for some students and a fervent one for others? Do we want to emphasize geometric and perceptual aspects of mathematics more for some students, and symbolic and algebraic aspects more for others? Do we want to make the learning of language more oriented toward the ear for some students and more toward the eye for those who have a poor sense of auditory discrimination? These are issues that have as yet scarcely been explored in educational philosophy or in discussions of educational policy. With the advent of the new technology they will become practical questions of considerable moment.

The third and final issue I wish to discuss is the place of individuality and human freedom in the modern technology. The crudest form of opposition to widespread use of technology in education and in other parts of our society is to claim that we face the real danger of men becoming slaves of machines. I feel strongly that the threat to human individuality and freedom in our society does not come from technology at all, but from another source that was well described by John Stuart Mill more than a hundred years ago. In discussing precisely this matter in his famous essay *On Liberty,* he said, the greatest difficulty to be encountered does not lie in the appreciation of means towards an acknowledged end, but in the indifference of persons in general to the end itself. If it were felt that the free development of individuality is one of the leading essentials of well-being; that it is not only a co-ordinate element with all that is designated by the terms civilization, instruction, education, culture, but is itself a necessary part and condition of all those things; there would be no danger that liberty should be undervalued, and the adjustment of the boundaries between it and social control would present no extraordinary difficulty.

Just as books freed serious students from the tyranny of overly simple methods of oral recitation, so computers can free students from the drudgery of doing exactly similar tasks unadjusted and untailored to their individual needs. As in the case of other parts of our society, our new and wondrous technology is there for beneficial use. It is our problem to learn how to use it well. When a child of six begins to learn in school under the direction of a teacher, he hardly has a concept of a free intelligence able to reach objective knowledge of the world. He depends heavily upon every word and gesture of the teacher to guide his own reactions and responses. This intellectual weaning of children is a complicated process that we do not yet manage or understand very well. There are too many adults among us who are not able to express their own feelings or to reach their own judgments. I would claim that the wise use of technology and science, particularly in education, presents a major opportunity and challenge. I do not want to claim that we know very much yet about how to realize the full potential of human beings; but I do not doubt that we can use our modern instruments to reduce the personal tyranny of one individual over another, wherever that tyranny depends upon ignorance.

The Schools

ANTHONY G. OETTINGER
Harvard University

Will the wonders of modern technology save our children from death at an early age? To pick up almost any current magazine or newspaper or to listen to eminent researchers and educational spokesmen is to be persuaded that, thanks to modern technology, the necessary educational revolution is just around the corner. But is it?

I share with many of my fellow computer scientists and engineers a solid faith that computers will ultimately influence the evolution of human thought as profoundly as has writing. I sympathize with hopes for programed instruction, language laboratories, television, and film in strips, loops, or reels. Ultimately, however, is not tomorrow. Education's institutional rigidity combined with infant technology's erratic behavior preclude really significant progress in the next decade, if significant progress is interpreted as widespread and *meaningful* adoption, integration, and use of technological devices (including books and blackboards) within the schools.

But the need to convince obtuse laymen, particularly the President and Congress, of the value of basic research often leads both scientists and educators into exaggerations that begin as well intentioned rhetorical devices but may end in self-delusion. Expediency leads politicians into their own scientific or technological promises in order to help pass a bill, win an election, or sugarcoat the bitter political pill of social reform.

In early March, I attended a conference on "An Educational System for the Seventies" (ES '70) convened by the U.S. Office of Education. One preconference announcement said:

It is hoped that the conference will serve these objectives: 1) to get consultative thinking from various groups about priority goals and outcomes for

Reprinted from Anthony G. Oettinger, "The Myths of Educational Technology," *Saturday Review,* 51 (May 18, 1968), pp. 76–77 and 91, copyright Saturday Review, Inc. Used with permission of the author and Saturday Review, Inc.

ES '70 in their subject matter area; 2) to provide practice in articulating desired student outcomes in terms of behaviors, values, attitudes, transfer to life situations, citizen role; 3) to provide cross-group communication and efforts at integrative thinking in exploring the realities of the organic curriculum; 4) to provide a limited yet critical exposure of the organic curriculum to secondary school leadership, teachers, and policymaking citizen groups.

It was quite clear on the first day of the conference that many participants believed that "ES '70" (note the noun) and "the organic curriculum" (note the noun phrase and the definite article) had a solid existence. This pervasive illusion was reinforced by a document accompanying the statement of objectives, which declared:

> This overall plan, the first phase of which is almost completed, will identify all of the activities that must be completed before the total new curriculum can become operational. These activities can roughly be classified as research, development, or demonstration.

The same document began with the claim that:

> Various elements of the educational process—such as team teaching, programed instruction, flexible scheduling, computer-assisted teaching, and individualized curricula—have recently been examined by researchers and judged to be important additions to current practice.

A pre-conference report on secondary education in the United States expressed this theme as follows:

> Educational researchers have made significant findings about the learning process, curriculum innovation, and educational technology. It is distressing when one considers the tremendous time lag between the initial research findings and the implementation of these findings. Even with a rapid escalation of federal research funds for education, the return on this investment has been inconsequential. In short, it seems that a massive and radical redesign of the secondary education program is imperative. To bring this about, a coordinated planning and development effort, involving a variety of social institutions, is necessary.

Few observers of contemporary American education will quarrel with the need for reform, or with the conclusion that the educational establishment is almost ideally designed to resist change. Much depends, therefore, on whether one believes that the "important additions to current practice" or the "significant finds" are really at hand to support "a massive and radical redesign of the secondary education program."

That is why distinguishing between ultimate promise and immediate possibility becomes so vitally important. Otherwise, the conclusion that the more exotic forms of educational technology can *now* be little more than placebos for the ailing social organism will be interpreted as a denial of their potential value, thereby unwittingly reinforcing the argument against further investment for "inconsequential" returns.

The expressed distress is real enough. The President and the Congress set great store on education as a weapon of social reform. The Office of Education is consequently under great pressure to produce immediate results. But when a program must be successful by definition, the need for a good show often overwhelms scientific objectivity; after the curtain falls, little remains either of practical value or of added insight. It may be expedient politically, when poverty is "in," to seek support of educational technology on the ground that it will solve the problems of our inner cities and to use it as a Trojan horse for wheeling in needed reforms. If, however, this leads to demands for an immediate return on investment, and if failure to produce this return is both probable and verifiable, then the expedient is not really good strategy. When ideas that are promising as objects of research and honest experiment give birth, through artificial dissemination, to a brood of hysterical fads, there is the danger that angry reaction will dump out the egg with the shell.

Dealing with the mythology of systems analysis requires making a distinction as delicate as that between ultimate promise and immediate possibility. The myth of systems analysis holds that educational salvation lies in applying to education the planning and control techniques commonly believed to have been successful in the defense and aerospace industries. Advocating systems analysis as a panacea ranks with making the world safe for democracy, unconditional surrender, and massive retaliation as an experiment in delusion for political ends. Yet, not to believe in the usefulness of systems analysis is to deny the value of reason, common sense, and, indeed, the scientific method.

Systems analysis cannot be dismissed as modern gadgetry. Its best formulations are indistinguishable from descriptions of the scientific method and thus have roots reaching back through Roger Bacon to Aristotle and not, as some believe, just to the RAND Corporation. At its best, therefore, the systems approach can be used in conjunction with well developed and reliable research designs to solve problems far more satisfactorily than naked intuition. The mathematical methods of control systems theory, for example, are very effective tools for designing speed controls for engines or process controls for certain chemical plants or in finding the best trajectories for missiles.

But there is far less validity than wishful thinking in claims for the success of the systems approach in the design, management, and control of entire space and military systems, in spite of the repeated citations of these enterprises as paradigms for educational and other social systems. It is also easily demonstrable that the educational system is much more complicated than any system yet devised by the military, and that we have much less understanding of the former's component parts.

Moreover, to the extent that systems analysis is used in the Defense Department as an analytic tool rather than as an administrative club, its value depends on the possibility of doing what Charles Hitch, the man who developed it for former Secretary of Defense Robert S. McNamara, has described as "explicit, quantitative analysis, which is designed to maximize, or at least increase, the value of the objectives achieved by an organization, minus the values of the resources it uses." To this statement Dr. Hitch has added caveats that many witless disciples have apparently forgotten and which therefore bear repeating:

> However, there are risks and dangers as well as opportunities in the application of new management techniques—including the risk of discrediting the techniques, if one tries to move too far too fast. Although it did not appear easy at the time, there is no doubt in my mind that the Department of Defense, or much of it, is easier to program and to analyze quantitatively than many areas of civilian government. For example, it is certainly easier than the foreign affairs area. Quite apart from these difficulties, the substantive problems in other areas are different and new. In Defense, we had several hundred analysts at the RAND Corporation and elsewhere developing programs and systems analysis techniques for a decade before the department attempted any large-scale general application.

Although the U. S. Office of Education, among others, nowadays sets great store on the possibilities of systems analysis, a planning-programing-budgeting system (PPBS), cost effectiveness, and similar things by other names, the evidence suggests that the continuation of Hitch's comment is valid for education:

> No remotely similar preparatory effort has gone into any other governmental area, and the number of trained and skilled people is so limited that they are inevitably spread far thinner in other departments of government than they were and are in Defense.

In any case, asking *how or how well* is silly unless we know *what*. The "systems" label should therefore not be given too much significance: It can produce no miracles; you *can't* just feed it to the computer. Neither should it be ignored. Despite its limitations, taking the systems viewpoint—namely, agreeing in principle that it is better to think about a problem in its whole context than not—is the best available attitude toward any subject, especially one whose literature is characterized by the *idée fixe* (individualized instruction), the panacea (applying computers to education can bring a powerful new force to bear on the central-city problem), and the empty label (organic curriculum). Thoughtful and thorough engineering is always good practice.

At this point, everything required by "the instructional system" is still in the experimental stage. While classroom scheduling by computer is advertised as a *fait accompli,* this is true only in the rather restricted sense of assigning students to conventional classroom groups and insuring that the number of groups matches the number of available teachers, and that these groups and teachers fit into available classrooms. Typically, the whole operation takes place once a term. Merchandising this unpleasant and tedious task is clearly a worthwhile and useful accomplishment, which deserves wider acceptance. It must be recongnized, however, that this is a far cry from keeping track of individual students week by week, day by day, hour by hour, or minute by minute, and matching them in turn with resources themselves parceled out in smaller packages than teachers per semester or rooms per semester. Packaging individual students is more complex than packing screws for dime stores or wrapping a lamb chop in plastic for the supermarket meat counter.

The goals for education have been stated with such monumental vagueness, and yet with such colossal residues of disagreement, as to provide no useful guidance for any systematic systems design. Current educational talk and writing is all for individualized instruction. There are several plausible reasons for grasping at this straw. For instance, many psychologists now officially agree that there are individual differences in learning capabilities. There obviously is also increasing consciousness that contemporary education does not serve equally the needs and the interests of all groups in our society.

But what does "individualization" mean? Lawrence Stolurow gives his definition in terms of an interaction between students and a computer. The student's "characteristics are stored in a student data base which permits the system to interpret responses in a selective way. Responses are also related to performance expectations. On this basis, the instruction becomes individualized." This statement of a research goal (*not* a current reality) is very attractive, since there is evidence that students prefer undivided attention from a computer to neglect from a human being.

A case can also be made for a narrower notion of "individualization"—something like "personalizing" or "customizing"—namely, taking a mass-produced object and stamping it with gold initials or heaping chrome on fins to give the illusion of individual tailoring. This is the sense in which current experimental computer programs greet a pupil with "Good morning, Johnny," by filling the blank in "Good morning, —————" with the name he had to give to identify himself to the machine in the first place. This is more genteel than, "Do not fold, spindle, or mutilate!"; "Hey, you!"; or "Good to see you, 367-A-45096!" It is, however, just as superficial, even when randomly selected variations heighten the effect of spontaneity.

A more charitable interpretation says that individualizing means giving full scope to idiosyncrasy. Harold Benjamin, in his witty but profound lecture *The Cultivation of Idiosyncrasy*, points out that this interpretation raises a question "which a democratic society may ignore only at its deadly peril." The question is double-barreled: 1) How much uniformity does this society need for safety? and 2) how much deviation does this society require for progress?

Yet the proponents of individualization pay only the scantiest attention to the fundamental goal-setting and policy decisions inherent in Benjamin's query. They often ignore even the most elementary systems-design consequences of their belief in individualization.

If—as is true of all present computerized systems of "individualized" instruction, as well as of many others based on explicit definitions of "behavioral objectives" (BOs)—the intent is to instruct students in such a manner that all will achieve a final level of competency which meets (or surpasses) the same set of minimally acceptable performance criteria, the objective cannot be the cultivation of idiosyncrasy. It is, rather, what an industrial engineer might call mass production to narrow specifications with rigid quality control. Each pupil is free to go more or less rapidly exactly where he is told to go.

Our present and most pressing problem is the lack of an empirically validated theory of teaching, and, in fact, we even lack a useful set of empirically validated principles of instruction that could form the primatives of a theory of teaching. This is not to say that we lack teaching practices that are widely used. Rather, it is to say that choice among existing practices cannot be made from data demonstrating the greater effectiveness of one over another. This problem, combined with the cost of computer systems, makes it more likely that we will reject a useful idea than it is that we will accept a useless one.

Semantic perversion, therefore, tends to mask the fact that the techniques now being developed may have great value in training to very narrow and specific "behavioral objectives," but do not address themselves to the many broader but just as basic problems of education.Training to minimal competence in well defined skills is very important in a variety of military, industrial, and school settings. It is not, however, the whole of what the educational process should be.

What conclusions can one draw about a desirable course of action? There is a strong temptation, reinforced by the accepted ritual of constructive criticism, to end a critical essay with an upbeat note, a ray of hope, and, preferably, the proposal of a favorite scheme of one's own. The systems analyst, however, like the doctor or lawyer, owes it to the ideal of professional integrity to tell his client the truth as he sees it, not as the client would like to hear it. He may or may not have a useful prescription. Polio has been conquered, a mumps vaccine has just come out, but the common cold and cancer are still with us.

However wasteful in appearance, it fits my prejudices best to encourage as much diversity as possible—as many different paths, as many different outlooks, as many different experiments, as many different initiatives as we can afford once the demands of education have been balanced against those of other needs of our society. We should plan for the encouragement of pluralism and diversity, at least in technique.

Advocating diversity in goals raises even deeper questions, although the two are linked. If, for example, we were to supply individual tutors to children (the expense is currently comparable to that of computer consoles), how much freedom of action would it be "safe" to give to tutors? The computer at least can lend itself to safe, guaranteed uniformity in its individualization.

This point of view argues, for example, toward channeling educational resources through pupils and their parents rather than through the educational establishment, federal or local. It seems vital to encourage greater freedom of choice in a situation which, however diverse in appearance because of the existence of 27,000 school districts, in fact has dreary monotony.

Vesting all educational authority in the federal government makes no sense, but letting our schools continue as local monopolies perpetuates on the local level a crime we would not and do not tolerate nationally. One could visualize under such circumstances a situation where various public or private organizations might create national school systems operating local schools as branch offices or as franchises under contract with the local school board.

Giving pupils the option to go to a town-operated school, a school operated by a neighboring town, a school operated in the given town by one company or another, could encourage competitive initiative. Unfortunately, this kind of approach also raises the specters of support to parochial schools, and of racial or economic segregation.

There may be other alternatives that could provide a kind of large-scale evolutionary effect with enough units at stake to create a fair probability that lots of different paths will be taken, and that illuminating controversy will rage. A miracle may make this happen with the seventeen schools participating in the ES '70 venture. Whatever the setting for educational experimentation, it is vital in our still profound ignorance to shy away from rigid prescriptions of either goal or technique. There is too much rigidity even in the present innovation fad which, ironically, diverts human and financial resources from both basic research and sustained application and evaluation efforts into the most visible quickie approaches that can sustain the illusion of progress.

Thought Processes

GEORGE A. MILLER

Rockefeller University

Some questions are like a cavity in a tooth; we keep coming back to probe them over and over until our tongues grow raw on their jagged edges. My topic is one of these. It has been explored almost without intermission for three hundred years. No one could estimate how many learned essays and lectures have been devoted to it. Dozens of articles and books sharing the generic titles, *Minds, Machines, and Other Things* are appearing daily. The pace at which these works are produced has grown in direct ratio to the complexity of the machines we can construct, which is to say that there has been an enormous outpouring of them in the last decade or two. The irritant for this recent outbreak of probing, of course, has been the emergence of automatic computing machines as a major influence in our lives. These new machines have enormously enlarged our conception of what a machine can be and do, and with every such enlargement it seems necessary to consider once again the ancient problem of the relation between men and machines.

Like most problems of any real importance, the relation between men and machines raises both theoretical and practical issues. The theoretical question —which is at least as old as the philosophy of Descartes—concerns the extent to which our brains can be considered as machines. The practical question— which is not entirely unrelated, but which, being practical, seems more immediately urgent—concerns the impact of these new machines on the social and economic institutions that regulate our daily co-existence. I shall consider the theoretical question first, because I think it has something to teach us that will be useful when we turn to more practical matters.

For many years I have studied the psychological processes that are entailed by our linguistic skills in communicating with one another, skills of

Reprinted from George A. Miller, "Thinking Machines: Myths and Actualities," *The Public Interest,* 2 (Winter 1966), pp. 92–97, 99–108, © by National Affairs, Inc. Used with permission of the author and National Affairs, Inc.

enormous complexity and uniquely human in character. Since my interest in the psychological aspects of communication is even older than the automatic computers, I can remember what those days Before Computers were like. When I try to compare them with the present, I can think of no summary statement more appropriate than that made by a famous American athlete who said, "I've been rich, and I've been poor, and believe me, rich is better." Believe me, computers are better.

Lest I confuse the reader with the puzzle of what computers have to do with the psychology of communication, however, let me plunge *in medias res*.

CAN MACHINES THINK?

Several years ago the English mathematician, A.M. Turing, considered the difficult question of whether or not a computing machine can think. Since the semantic and metaphysical issues involved are apparently unresolvable, Turing rephrased it. Can a computing machine, he asked, behave in the way we behave when we say we are thinking? This rephrased question, he felt, might have an answer, and to make the issue perfectly definite, he proposed what he called the "imitation game." In the imitation game a computer is compared with a human being in terms of the answers it gives to an interrogator; if the interrogator is unable to determine when he is communicating with a human being and when he is communicating with a computer, then, Turing would say, the machine must be behaving in a human manner. And that, he implied, is all anyone should ever mean by the question, "Can machines think?" Turing, writing in 1950, predicted that within fifty years it would be possible to build and program computers that could do well at the imitation game. We have not seen them yet, but his prophecy has several years to run.

I have referred to Turing's question not because I wanted to approve or disapprove of it—certainly not because I think I can answer it—but rather because it illustrates how intimate are the relations among computers, communication, and cognition. To understand better the cognitive processes we call thinking, Turing proposed to simulate them on a computing machine and to test the quality of the simulation in terms of the machine's performance in a communication situation. Human intelligence is best demonstrated when we communicate; if a machine is to be considered our equal, then it must communicate as we do. Which illustrates how such apparently unrelated topics as computers and communication can become important for psychologists.

Two easily recognizable groups frequently object when this question is seriously discussed. One group feels that such a question is morally reprehensible, that to compare man and machine diminishes the human spirit. Of course, science has been whittling away at our self conceit ever since it pushed us out of the centre of the universe and discovered the apes in our family tree, but whether science has thereby dimished or augmented human dignity is not entirely clear. In any case, these are not the critics I wish to answer. I respect their opinions; I hope they respond in kind.

In many ways the second group is more interesting, for it consists of men who know computers thoroughly, from top to bottom and inside out. Many of the real professionals—men who developed these wonderful machines and discovered how to use them—consider the question absolutely absurd. They understand all too well the limitations of their new toy and they would blush crimson if anyone caught them referring to it as a "giant brain" or a "thinking machine." They know whose brains did the real thinking behind all this new technology, and it was not the machine's.

I feel these objectors must be taken seriously, for they have earned their right to respect from those of us who hope to profit from the interaction of computers, communication, and cognition. The nature of their objections can be revealed most clearly, I believe, if we review briefly something of the history of their work.

THE AUTOMATIC DESK CALCULATOR

Leaving aside the well known story of Charles Babbage and his Analytical Engines, the history of modern digital computers began about twenty-five or thirty years ago when engineers attempted to make the ordinary, manually-operated desk calculator fully automatic.

Think for a moment of the way a human operator uses an adding machine. He begins with numerical data and a formula into which they are to be inserted. From the formula he sets up a sequence of operations that must be performed. Following this sequence, he pushes keys to put numbers into the machine, then pushes other keys to tell the machine what arithmetical operations to perform, and finally copies down the result on a piece of paper. These results can then be put back into the machine and further operations performed on them in turn, and the cycle repeated until the full list of instructions has been executed.

Key punching and copying are slow and tedious, so it is natural to think that the machine might just as well do them for itself. This was the idea behind the first "fully automatic" computers. The complete sequence of instructions was prepared in advance in a form the machine could sense and interpret; usually in the form of holes punched in a long paper tape. All the data were similarly prepared in advance on another tape. Then, instead of requiring the operator to copy down the intermediate results and feed them back into the machine, the machine was given a memory of its own—a sort of mechanical scratch pad—where numbers could be stored temporarily until they were needed. Once all was ready, the operator simply pressed a button and the whole computation ran off automatically. Not only did this enlarged adding machine eliminate most of the mistakes that human operators seem unable to avoid; it also worked much faster, so that computations previously considered too laborious to undertake by hand could now be accomplished in a few hours.

All this is a familiar story, of course. I mention it only to recollect an attitude toward computers that prevailed in those days. Recall the use made of the early machines. One of the first projects was to compute the values of various important mathematical functions with great accuracy and to publish the resulting tables in order to make them available to scientists and mathematicians. No example could better illustrate how completely they missed the significance of their own invention. They wanted to make it easier to perform accurate computations in the traditional sense of that term, and they knew from personal experience how valuable good mathematical tables had always been to a working mathematician. What they failed to see was that the computer itself made the mathematical tables unnecessary. From that time on nobody would bother to refer to a table he could simply ask a computer to generate the value of the function as needed in the course of a computation.

THE AUTOMATIC FILING CABINET

It was, of course, this initial focus on numerical calculation that gave the new machines the name "computers." If that name had never been adopted, if we were suddenly faced for the first time with the modern machines and asked to find an appropriate name for them, I doubt that "computer" would now be our first choice. "Information processing machines" would be more likely. Computing is only one of many operations a contemporary machine can perform, and some of its applications are wholly nonnumerical. But it was originally conceived as an extension and enlargement of the adding machine, so "computer" it has been ever since. When you look at computers as glorified adding machines, of course, there is little temptation to claim that they are thinking any more than their smaller ancestors had been thinking while their gears spun around.

Development did not stop there, however. If the first generation of computers can be said to have been modelled on the idea of the adding machine, the metaphor that is most appropriate for the second generation is the filing cabinet. I said that a certain amount of memory had been provided in the first machines, enough to enable them to store intermediate results temporarily during the process of computation. In the next stage this feature of the computer expanded enormously. All the skill and ingenuity of the inventors and machine designers was directed toward enlarging the capacity for storing information.

One consequence was to increase the speed at which machines could operate. To consult punched tapes every time a new datum or a new instruction was needed is relatively slow. With an enlarged memory, both data and instructions could be stored in the computer before the computations began, and could be retrieved with the speed of electrical conduction. As John von Neumann foresaw, storing the program of instructions in the machine turned out to be an especially significant advance, because a machine could then modify its own instructions as the computation proceeded and could select its next instruction from any point in the program depending on the outcome of preceding instructions. The flexibility of programming that resulted from this simple but profound innovation is an essential characteristic of modern digital machines.

Not only did larger memories make computation more efficient; they also made possible new applications for the machines in business and government. The contents of the filing cabinets—inventories, accounts, personnel records, and all varieties of economic and statistical information—could be dumped into the computer's enlarged memory, there to be processed by the fastest and most accurate bookkeeper ever created. And so it came to pass that the development of computers with large memories rewarded the customer and manufacturer alike. Without this financial support, the development of the new machines would never have been economically feasible.

The addition of a large memory, however, still did not turn a computing machine into a thinking machine. If a desk calculator does not think, and if a filing cabinet does not think, why should anyone imagine that they would start thinking when we put them together?

THE AUTOMATIC VOLTMETER

The third generation of computing machines also had a metaphorical progenitor: the voltmeter. The simplest kind of voltmeter has a pair of electrodes that serve as its sense organs to pick up differences in electrical potential, and some kind of scale-and-pointer arrangement to publish the measured voltage. When engineers began to dream about the possible embellishments they could add to this simple scheme, a whole sequence of new devices began to appear. First came the substitution of a cathode-ray oscilloscope for the scale-and-pointer. This resulted in an extremely useful instrument that converted electrical voltages into visible wave forms. A certain amount of electronic circuitry is needed to generate this visual display, of course, but not so much that it cannot be built into an easily portable cabinet. But as computers entered the scene it was inevitable that they would be used to process the incoming electrical data in ever more complicated ways before it was displayed. The result was the development of the most flexible and intricate "voltmeters" that the fertile engineering mind could imagine. This time economic support came from the military departments, because such systems are extremely valuable for processing information from radar receivers and other sources and presenting it in forms most convenient for military commanders.

For these computers the data does not have to be collected in advance and painstakingly copied in a form acceptable to the computer. These new machines have their own sense organs and can feed information directly into the computer, untouched by human hands, where it can be digested processed, and displayed in a form more intelligible to a human operator. Their usefulness is not limited to military systems, of course. Wherever computations must be performed in what engineers like to call "real time," wherever an immediate display of the processed data is required, these supervoltmeters prove their value.

Take an adding machine, give it access to a filing cabinet, attach a battery of sensors to report events in the environment, throw in an oscilloscope or two for instantaneous communication with the operator, and you have a very modern computer. In some respects its general design resembles an organism's, which is the reason we use such terms as "memory" and "sense organs" to describe it. But those engineers who watched it grow, who know how much real human thinking is required to compose its instruction programs, are still not inclined to speak of it as a "thinking machine." . . .

GIANT BRAINS

Computers modelled after adding machines, filing cabinets, voltmeters—these are all well and good. But how long will it be before we have computers modelled after the human brain? Some believe we have already accomplished it. To balance the enthusiasts are the sceptics who believe we can never accomplish it.

It seems perfectly obvious to me that both parties are speaking beside the point. Whatever else a brain may be it most certainly is *not* a digital computer Until we begin to develop computers along entirely different lines than we have in the past, I see little hope of finding anything but crude analogies between them. We are today about as far from building a computer modelled on the human brain as Archimedes was from building an atom bomb. I do not mean this as a criticism of computer engineers. I mean simply that we do not yet know enough about the brain or the principles on which it operates.

In the first blush of enthusiasm over the new machines we heard a great deal about analogies between relays and synapses, between electrical pulses and the nerve impulse, between the wiring of a computer and networks of neurons in the brain, etc. Some scientists still talk this way, but I believe that most people who have seriously compared the computer to the brain are more impressed by the differences than by the similarities.

You can build a computer along the lines of an adding machine, or a filing cabinet, or a voltmeter because these devices are relatively simple and because, since we invented them, we can understand them. The situation is quite different when we talk about building a computer modelled on the living brain. A brain is not some simple gadget that we conceived and built; it operates in ways still unknown to science or technology. Until its general principles are understood, it is vain to talk of building computers based on them. Let us, therefore, put aside the notion that there should be any point-for-point resemblance between computers and brains. We are not—or should not be—concerned with superficial analogies.

Which seems to put a full stop at the end of one line of speculation.

THE CYBERNETIC APPROACH

Why, then, should a psychologist, or any other student of living organisms, find computers interesting? And it is a fact that many of us do find them so. It is not merely because they process our data for us, or solve our equations for us, or control our experiments for us, although these services are all valuable. Our enthusiasm comes from a different source.

Someone has pointed out that a major difference between the physical and the biological sciences lies in the fact that the physical scientist formulates propositions of greater generality. A biologist is properly concerned only with organisms that actually exist. A physical scientist, however, is free to consider the set of all possible universes, of which our actual universe is only a special instance. This difference in method grows a bit hazy in the field of molecular biology, where biologists have adopted most effectively the strategy of the physical scientists, but at the level of whole, intact organisms adapting to their natural and social environments, I consider the difference both real and significant. Physicists deal generally with both the actual and the possible; biologists are largely confined to the realm of the actual.

Suppose, however, that we were to approach the study of whole organisms from a physical point of view. Suppose, that is, we were to study not merely the organisms that actually exist, but were to consider the full class of all possible systems—whether they exist or not, whether they are animate or inanimate—that might perform functions biological systems are known to perform. If this abstract approach were feasible, we might hope to formulate theories so universal and so powerful that they could be applied to organisms and machines alike. Then we would not be entangled in a fruitless argument that organisms are nothing but machines. Instead, both machines and organisms, insofar as they performed the same function, would be seen as particular instances of theoretical systems of far greater generality.

It is this possibility that provides the real source of our excitement, and keeps the cyberneticist hard at work in the face of all criticism that actual organisms and actual computing machines are very different things. He hopes to look beyond these actual instances to discover general principles governing all possible systems.

In order to adopt this abstract approach it is essential to select some clear and well-defined function as our starting point. We must decide in advance what function we wish to generalize and then concentrate on defining and reproducing that aspect of behavior at the expense of all others. If, as in Turing's imitation game, we decide it is the function of linguistic communication we are going to generalize, then we will probably forget about the other functions—digestion, say, or driving an automobile—that people also perform. (Turing carried this abstraction even further; he was willing to let the interrogator communicate with his human and mechanical partners by teletypewriter, thus leaving out of account the problem of producing natural speech with all its intonations, hesitations, and subtle shadings.) Turing eliminated these other aspects of normal human behavior both to simplify and to clarify the problem; other aspects might be simulated as well, but the significantly human accomplishment that lies at the heart of his problem is our ability to string words together in meaningful, grammatical sentences. He abstracted one important and reasonably well defined function that people perform and posed the problem of generalizing that particular function.

Our aim, therefore, is to enunciate general principles of the following form: "If any device is to perform function X, then that device is subject to or limited by the principles Y which must hold for all possible devices performing this function."

We want to formulate general principles that will describe all possible devices of a given class, regardless of their particular anatomy or mechanism.

Now, with this aim in mind, consider one strategy we might adopt in our search for such general principles. It is obvious that machines have taken over many functions previously performed only by human beings. If we examine all the different ways in which machines have accomplished any one of these functions, we should find—since we understand the machines quite well—that they have certain features in common. If we can show that these common features are not the result of some poverty in their inventor's resouces or imagination, but follow necessarily from the nature of the function being performed, then we know that they must also apply to human beings insofar as human beings are also able to perform that function.

More often, however, we will find that our mechanical solutions fall into several distinct types, in which case we may be able to say that any device performing function X must be of type A, or type B, or type C, etc. Then we know that a human being, insofar as he also performs function X, must be of type A, B, or C, etc. We are then faced with a well-defined empirical problem. Can we determine which type a living organism is? At this point, an experimental scientist must undertake to devise tests that will settle the question.

Unfortunately, we are not always in the position of knowing that all devices performing a particular function *must* be of a single type, or *must* be one of a limited number of possible types. Even after we have invented three or four ways to perform the function, we may still be unable to show that we have exhausted all the possibilities. There may be other solutions that we have not been clever enough to see. This situation is far less satisfactory, but it is still worthwhile to try to determine whether or not living organisms belong to one of the general types that we have invented. If the answer is "yes," then we understand that much more about the living organism; if the answer is "no," then we are encouraged to continue our search for alternative ways of performing the function by machines. In either case, we know more than we did before.

These remarks on the virtues of abstraction have, I fear, been quite abstract in themselves. So let me resort to a few examples of this strategy of research.

AN EXAMPLE: NOISE

First, the case in which all devices performing function X must be of a single type. One example would be the following: any device that performs the function of a communication channel can produce only a finite number of distinguishably different output signals per second. Even in the best communication systems there is a residual level of noise that cannot be eliminated; as we try to make finer and finer distinctions between the output signals, we will eventually encounter this random noise, which sets a limit to the accuracy of our discriminations. This generalization must hold for human beings as well, insofar as human beings can perform the function of a communication channel. Offhand, this principle is so general that there would seem to be little for an experimental scientist to do about it. In fact, however, it has led to a great deal of experimentation by psychologists, who have wanted to measure the noise level of the human channel. The measurement is generally stated in units called "bits" of information. It has been estimated that, under optimal conditions, a human being has a channel capacity of about 25 bits per second. This means that each second a human channel can select any one from about $2^{25} = 200,000,000$ distinguishably different responses. This may impress you as a very large number, but let me remind you that our electronic communication channels regularly transmit thousands or millions of bits of information per second. By comparison, our capacity of 25 bits per second is puny indeed. Considered as communication channels, human beings have a rather high noise level.

I would like to call attention to the exact wording I have used. I said that this general law must hold for human beings *insofar as human beings perform the particular function* in question. What I have *not* said is that, since the laws governing communication channels can be applied to human beings, human beings are nothing but communication channels. I always distrust the man who says that human beings are "nothing but" something else, for he is deliberately concealing the abstraction on which his claim is based. Human beings are nothing but human beings.

AN EXAMPLE: SUBROUTINES

Next, the case in which all devices performing function X must belong to one of a limited set of possible types. Let us begin with a simple example.

Consider the following: any device that is to perform the function of answering a question must either (*a*) obtain the answer by consulting some memory where the answer is stored, or (*b*) synthesize the answer from other information according to a set of rules. If for example, you are asked the value of the logarithm of 75, you can either (*a*) remember it—which would include looking it up in a table that remembered it for you—or (*b*) compute it by carrying out some rather tedious calculation that most of us do not remember how to do. Similarly, if in the course of solving some problem on a computer it will be necessary to know the value of a logarithm, we either store a table of logarithms in the computer's memory in advance, or we must give the computer a sequence of instructions that will enable it to calculate the value when it is needed. With a computer, alternative (*b*) would generally be preferred, since it uses the memory more efficiently, and we would prepare what is called a "subroutine" for computing logarithms. Each time the value of a logarithm was needed in the course of executing the main routine, the computer would interrupt what it was doing, refer to the logarithm subroutine to calculate the answer, and then resume the main routine where it had been left off.

In most cases both techniques are employed; some information is stored in memory, other information is reconstructed as needed. A fascinating problem for a psychologist is to try tease apart these two methods of producing answers in human beings. The question has to be asked about each area of information separately, but it is my strong impression that we, like the computer, make very extensive use of the reconstructive method, that we remember most things by following rules for deducing what we need to know from other facts. We are, in short, equipped with a large assortment of subroutines —using that term now in a broad sense—that we can use to generate answers as they are needed.

AN EXAMPLE: RECURSIVE SUBROUTINES

For a more esoteric illustration, let me refer to some of my own research. The idea behind it came from a comparison of the structure of computer programs and the structure of grammatical sentences, so I must introduce it with a few remarks about the way such structures are put together.

Once we have reached the level of complexity in programing computers where we can have subroutines stored away for use as needed, an interesting possibility emerges. Ordinarily, we interrupt the main routine to perform the subroutine, then return to the main routine when the subroutine is finished. It is possible, however, for one subroutine to refer to another subroutine. That is to say, we can program a computer in such a way that the subroutine itself is interrupted while some other subroutine is executed; when this second subroutine is completed, the computer returns to the first subroutine again, and when it in turn is finished, goes back to the main routine. The interruption is itself interrupted. Any busy person will recognize how easily this can happen.

An interesting situation arises when we ask whether or not a subroutine can refer to itself. Consider what this would mean. The computer interrupts its main routine to go into subroutine S. In the middle of executing subroutine S, however, before S is finished, the computer is instructed to stop what it is doing and begin to execute subroutine S all over again. This may sound a bit complicated, perhaps, but actually it is not. No trouble will arise until subroutine S is completed and the computer must decide where to resume its work. Unless special precautions are taken in writing the program, the computer will not be able to remember that the first time it finishes subroutine S it must re-enter subroutine S, and the second time it finishes subroutine S it must return to the main routine. In the slang of computer programmers, the second re-entry address is likely to "clobber" the machine's memory of the original re-entry address. The situation gets even more tangled, of course, if subroutine S can call on itself repeatedly, for each time it does so a new re-entry address must be remembered.

Programs that are written in such a way that subroutines can refer to themselves repeatedly in mid-flight are generally called "recursive," and programs that do not make such provisions are called "non-cursive." This gives us our two possible types of programs, or, if you like, two possible types of systems. It turns out, for reasons I will not go into, that recursive systems are intrinsically more powerful than non-recursive systems; they can do everything a non-recursive system can do, plus some other things that are impossible for the non-recursive system. So the distinction I have described, although it may seem rather subtle, is very important. Recursiveness is a desirable property to have in a computer programming language, and many ingenious stratagems have been devised to make it available.

Now, enter the psychologist, armed with a conviction that people use subroutines in their cognitive operations. Since any device that consults subroutines is either recursive on non-recursive, and since people use subroutines, which type of device are they?

One way to investigate this question presents itself in the realm of language. It is a feature of natural language—by "natural language" I mean the languages we ordinarily use in speaking to one another, as opposed to the "artificial language" that we have developed for mathematics, logic, computer programming, and so on—it is a feature of natural languages that sentences can be inserted inside of sentences. For example, *The king who said, "My kingdom for a horse," is dead,* contains the sentence, *My kingdom for a horse,* embedded in the middle of another sentence, *The king is dead.*

Think of a listener as processing information in order to understand this sentence. Obviously, his analysis of one sentence must be interrupted while he analyzes the embedded sentence. When he finishes analyzing the embedded sentence, he must then resume his analysis of the original sentence. Here we have all the elements present in a computer subroutine.

The question, of course, is whether we can do this more than once, that is to say, recursively. Let us try: *The person who cited, "The king who said, 'My kingdom for a horse,' is dead," as an example is a psychologist.* Most people find this on the borderline of intelligibility: if I had not prepared you for it, you probably would not have understood. Let us go one step more: *The audience who just read, "The person who cited, 'The king who said, "My kingdom for a horse," is dead,' as an example is a psychologist," is very patient.* By now you should be ready to give up. If not, of course, I could go on this way indefinitely.

A PSYCHOLOGICAL EXPERIMENT

Even though they are grammatical, such sentences are obviously difficult to understand, which suggests that our ability to use subroutines that refer to themselves must be rather limited. My colleagues and I, however, wished to make a more objective measurement, so we asked people—students at Harvard University—to memorize sentences with various amounts of self-embedding in them. The grammatical device we used to embed sentences inside of sentences was the relative clause, which is particularly convenient because all of the sentences can have the same length.

Let me build up one example for you. Begin with the following five sentences: *The movie was applauded by the critics, The script made the movie, The novel became the script, The producer discovered the novel,* and

She thanked the producer. The most intelligible way to combine these five sentences into one involves no embedding at all:

She thanked the producer who discovered the novel that became the script that made the movie that was applanded by the critics.

There is nothing difficult here. The structure of this sentence is the same as the nursery rhyme, "This is the cow with the crumpled horn that tossed the dog that worried the cat that killed the rat that ate that malt that lay in the house that Jack built." And that, of course, is easily undertsood and enjoyed by young children.

Now let us embed one of these relative clauses:

The producer (whom she thanked) discovered the novel that became the script that made the movie that was applauded by the critics.

There are still no problems. Everyone who knows English can drop into a subroutine for analyzing one embedded relative clause.

The interesting situation arises when we insert another relative clause into the middle of the first one, as follows:

The novel (that the producer (whom she thanked) discovered) became the script that made the movie that was applauded by the critics.

Now the plot begins to thicken, and it gets even thicker when we do it once more:

The script (that the novel (that the producer (whom she thanked) discovered) became) made the movie that was applauded by the critics.

Finally, with four embeddings:

The movie (that the script (that the novel (that the producer (whom she thanked) discovered) became) made) was applauded by the critics.

If you are able to understand this final version, it is only because I led you into it gradually. Our students did not have such a gentle initiation. They would hear for the first time something like:

The story that the book that the man whom the girl that Jack kissed met wrote told was about a nuclear war.

Their task was to memorize it, and we measured how well they could remember it as a function of the amount of studying they had done. Every version of the sentence contained exactly the same 22 words. All we did was to rearrange their order a bit, and so rearranged the order in which the listener's cognitive operations of analysis had to be performed. In spite of their unusual appearance, however, all of these embedded sentences are perfectly grammatical, by any reasonable interpretation of English grammar.

I think you can predict from your own reactions the performance of the subjects in our experiment. The simplest way to summarize the results is to say that everyone could handle one embedded clause, some could handle two, but everyone had trouble with three or more. The ability of some people to handle two embeddings indicates that we are not entirely bereft of recursive facilities, but their inability to deal easily with three or more tells us that our recursive resources, whatever they may be, are extremely limited, even in subjects as intelligent as Harvard students are reputed to be.

Introspection—that unreliable but irresistibly convenient tool of the psychologist—indicates that all is proceeding quite well with the embedded sentence until we encounter the long string of predicates, "...thanked discovered became made was applauded . . . ," at which point our grasp of sentence structure collapses and we are left with a haphazard string of verbs. We are unable to locate the subjects associated with each successive predicate and that, of course, is exactly what we would predict if people were analyzing them as would a non-recursive computing machine that could not remember its re-entry addresses. This subjective hunch has been tested objectively by studying eye movements as people try to read such embedded sentences. Their eyes move forward along the line in a normal fashion until they come to the third or fourth verb; at that point, regressive eye movements occur as they begin to look frantically back and forth for the subject associated with each verb.

Now, I would be the last to claim that this little experiment solves all the problems of psychology, but I do think it is amusing and that it establishes an important point, namely, that we are very poor at dealing with recursive interuptions. If this result is confirmed in studies of interruption in other kinds of tasks, we may have to assign human beings to the general category of non-recursive devices. The fact that we are able to process information as effectively as we do without this powerful tool makes our cognitive functions all the more fascinating as objects for scientific research.

It is true that in everyday affairs we do not seem to suffer too severely from this limitation. But if you will recall the everyday situations in which you were able to resume what you were doing after your interruptions had been interrupted, I think you will agree that you were able to resume because the interrupted task itself remembered for you. If you are interrupted while painting a wall, when you return the wall will provide an unmistakable reminder of how far you had gone and where you should resume. It is only when you cannot count on the environment to remember your re-entry point that your cognitive limitations become a handicap. Perhaps it is because we can usually count on the task to remember for us that we have not evolved more extensive powers for recursion.

When we know that living organisms must, insofar as they are able to perform function X, fall into one of a limited set of types of devices for performing that task, we are in a relatively good position to learn something interesting about them. In most situations, however, the most we can say is that there are several different ways a machine might perform the function, and all we can ask is whether the organism performs it similarly. If the organism does not—which will usually be the case unless we are very lucky—there is little we can do but continue to study the problem. There are so many examples of this sort, each surrounded by its own special penumbra of ignorance, that I dare not launch into examples.

IMPLICATIONS FOR THE FUTURE

I trust these examples have managed to convey some sense of the detail with which a function must be analyzed before we can begin to talk of performing it with a computing machine. Learning to cope with the extreme literalness of computers is good discipline for a psychological theorist, for many of us are inclined to rely implicitly on common-sense explanations that tempt us to think we understand a process when, in fact, we cannot describe any detailed operations by which it might be accomplished.

As we have begun to spell out in detail what these cognitive operations might be, we have begun to see that above—or perhaps behind—the mass of detail there are often very general principles that govern the operation of any device, living or non-living, capable of performing the function in question. It is not a matter of reducing men to machines, but of discovering general principles applicable to men and machines alike. And this is an exciting prospect.

Practical applications of this kind of knowledge are difficult to foresee with either confidence or clarity, but I believe I can point to one general consequence that will emerge as our knowledge of information processing systems increases. By classifying man ever more accurately with respect to his capacities and incapacities for processing information, by discovering more about the general system that he exemplifies, we will gain increasingly deeper insight into how best to use computers to perform functions that are difficult for him. As our understanding increases, I think we will be better able to optimize the man-machine team. Mechanical intelligence will not ultimately replace human intelligence, but rather, by complementing our human intelligence, will supplement and amplify it. We will learn to supply by mechanical organs those functions that natural evolution has failed to provide.

Those of us who are optimistic about this general strategy of research expect that it will prove valuable in all areas of the biological, psychological, and social sciences. Perhaps we should restrain our enthusiasm until we have more substantial accomplishments to report. But if these advances actually occur, they will undoubtedly have rather profound effects on our lives as ordinary citizens that may entail major readjustment in our conception of ourselves and our social institutions. If such adjustments do lie ahead, as a consequence of our advancing knowledge of information processing systems, it would be negligent of scientists not to discuss them publicly.

Thought Processes

R. W. HAMMING

Bell Telephone Laboratories

One of the most common questions asked is, "Can computers think?" The answer, of course, depends on the precise definition of "thinking," which practically no one is prepared to define sharply.

Let us examine some of the available data. Machines have been "programmed" to play such thinking games as chess and checkers, and all things considered, they play fairly well—in checkers well enough to beat many human players. They have also been programmed to solve high school geometry problems. By "solve" and "play a game" we do not mean the machine consults some large table and selects the correct move. Rather, we mean some human or humans have described a method by which such problems can be approached—usually with no guarantee of success. The methods are not necessarily those which humans themselves use, but they tend to resemble our rationalizations of how we solve problems and play games.

Inevitably the question comes up, "Can a computer produce a new result?" Certainly machines have produced results which surprised the humans who planned the program. To take one classic example, the geometry proving routine was asked to prove that in a triangle ABC if AB equals AC then angle ABC equals angle ACB—the well-known theorem that if two sides of a triangle are equal then the base angles are equal. Most humans faced by this problem either bisect the angle A and produce two congruent triangles, or else draw a line from the vertex A to the midpoint of the base BC, again producing two congruent triangles. The machine routine merely observed that triangle ABC was congruent to triangle ACB and hence corresponding angles were equal. The proof is both short and elegant. It was

Reprinted from R. W. Hamming, "Intellectual Implications Of The Computer Revolution," *The American Mathematical Monthly*, 70 (January 1963), pp. 9–11, by permission of the author and The Mathematical Association of America.

231

known to Pappus.* But if you were to examine the routine the machine used you would probably find that the machine was programmed first to see if it could prove the theorem, and if not then try to add a suitable line and try again. The result is, then, easily explained—the machine did what it was told to do. But then are we so different? Were we not programmed, haphazardly to be sure, to solve problems? What was our high school course in geometry all about except to program us to solve problems?

Thinking is closely associated with learning, and perhaps you feel that the crux of the matter is self-improvement. Consider, then, the fact that I could take two copies of a chess playing program, put them both in the same computer, but with one coefficient in one program changed. I could then let the machine play one formula against the other. Due to the fact that we almost always include in a game playing program a random choice to be used when two or more moves are rated as about equal, the machine will play, say, ten different games. Suppose one formula wins seven out of ten games. I could then have the machine continue to change the coefficient in the favorable direction until no further improvement was observed. In this fashion I could have the machine go through the coefficients one at a time until all of them had been improved. I would also probably try to change combinations of two at a time. Thus the machine from experience would improve the quality of its game. In evolutionary terms, it would be the survival of the fittest program.

We can imagine going further. Suppose I had a collection of small strategies. I could have the machine substitute whole pieces here and there into the program to see if the program were thereby improved. In the biological analogy, these are mutations to be selected for their survival by success or failure in competition with other programs.

Now that you can see survival of the fittest using both small variations as well as occasional large mutations, are you so sure that a program cannot produce "thinking"—whatever the word means?

Very frequently "thinking" is defined to be what Newton did when he discovered gravitation. By this definition most of us cannot think! As an exercise I suggest you try framing a test that is the least, or close to the least, which you will accept as demonstrating that a machine can think. I have been unable to devise one that would suit myself, let alone others, and have tentatively considered the hypothesis that "thinking" is not measured by what is produced, but rather is a property of the way something is done.

Before leaving the general area of thinking I wish to point out the fact that machines provide a fruitful approach to many questions. Back in 1939 Turing, a British logician, imagined a computing machine, now called a Turing machine, to prove some theorems in abstract logic. The machine was a paper machine in the sense that no actual machine was ever contemplated, rather the conception of a machine was used to aid in the analysis and proof of the results.

Such an approach has been used many times, but the presence of actual

* See Sir T.L. Heath, *Euclid's Elements* (New York: Dover Publications, Inc., 1956), p. 254.

machines has greatly stimulated the general field. Thought experiments are now fairly common in some fields. The discussion of chess playing I gave is an example of a thought experiment. To carry out the idea on an actual computer would be very expensive in money and time. I have also on several occasions examined a problem as if I were going to put it on a computer, though I had no intention of actually doing so, and in the examination found the answer I was looking for.

Such an approach requires you to give an absolutely complete description without skipping lightly over some detail that you think is obvious. As an example of oversight, years ago in a calculus class I taught a certain process called "integration by parts," yet when I now try to give a description to a machine I find that there are many details I do not understand well enough to write out a program for the machine. The students had the impression, along with me, that they understood the process, but they too probably cannot give a detailed description to a machine.

At times the machine approach can be very fruitful—and it can certainly pinpoint obscurities very rapidly as well as expose ignorance. I note that increasingly in abstract books authors are appealing to a machine model for clarity of expression. I suggest, therefore, that the habit of asking for a machine description of something will become widespread wherever it is desired to know clearly what one is talking about. Without a detailed description in some language a machine can use there is no conviction that you know what you are talking about; with it there is the illusion you do.

The Sciences

ANTHONY G. OETTINGER

Harvard University

In its scientific applications the computer has been cast in two quite distinct but complementary roles: as an instrument and as an actor. Part of the success of the computer in both roles can be ascribed to purely economic factors. By lowering the effective cost of calculating compared with experimenting the computer has induced a shift toward calculation in many fields where once only experimentation and comparatively direct measurement were practical.

The computer's role as an instrument is by far the more clear-cut and firmly established of the two. It is in its other role, however, as an active participant in the development of scientific theories, that the computer promises to have its most profound impact on science. A physical theory expressed in the language of mathematics often becomes dynamic when it is rewritten as a computer program; one can explore its inner structure, confront it with experimental data and interpret its implications much more easily than when it is in static form. In disciplines where mathematics is not the prevailing mode of expression the language of computer programs serves increasingly as the language of science. I shall return to the subject of the dynamic expression of theory after considering the more familiar role of the computer as an instrument in experimental investigations.

The advance of science has been marked by a progressive and rapidly accelerating separation of observable phenomena from both common sensory experience and theoretically supported intuition. Anyone can make at least qualitative comparison of the forces required to break a matchstick and a steel bar. Comparing the force needed to ionize a hydrogen atom with the force that binds the hydrogen nucleus together is much more indirect, because

Reprinted from Anthony G. Oettinger, "The Uses of Computers in Science," *Scientific American,* 215 (September 1966), pp. 161–166, 168–170, 172, Copyright © by Scientific American, Inc. Used with the permission of the author and Scientific American, Inc.

the chain from phenomenon to observation to interpretation is much longer. It is by restoring the immediacy of sensory experience and by sharpening intuition that the computer is reshaping experimental analysis.

The role of the computer as a research instrument can be readily understood by considering the chain from raw observations to intuitively intelligible representations in the field of X-ray crystallography. The determination of the structure of the huge molecules of proteins is one of the most remarkable achievements of contemporary science. . . . The labor, care and expense lavished on the preparation of visual models of protein molecules testify to a strong need for intuitive aids in this field. The computational power required to analyze crystallographic data is so immense that the need for high-speed computers is beyond doubt.

The scope and boldness of recent experiments in X-ray crystallography have increased in direct proportion to increases in computer power. Although computers seem to be necessary for progress in this area, however, they are by no means sufficient. The success stories in the determination of protein structures have involved an interplay of theoretical insight, experimental technique and computational power. . . .

The metaphor of the transparent computer describes one of the principal aims of contemporary "software" engineering, the branch of information engineering concerned with developing the complex programs (software) required to turn an inert mound of apparatus (hardware) into a powerful instrument as easy to use as pen and paper. As anyone can testify who has waited a day or more for a conventional computing service to return his work only to find that a misplaced comma had kept the work from being done at all, instant transparency for all is not yet here. Nevertheless, the advances toward making computer languages congenial and expressive, toward making it easy to communicate with the machine and toward putting the machine at one's fingertips attest to the vigor of the pursuit of the transparent computer.

A few critics object to the principle of transparency because they fear that the primary consequence will be atrophy of the intellect. It is more likely that once interest in the *process* of determining molecular structure becomes subordinate to interest in the molecule itself, the instrument will simply be accepted and intellectual challenge sought elsewhere. It is no more debasing, unromantic or unscientific in the 1960's to view a protein crystal through the display screen of a computer than it is to watch a paramecium through the eyepiece of a microscope. Few would wish to repeat the work of Christian Huygens each time they need to look at a microscope slide. In any case, computers are basically so flexible that nothing but opaque design or poor engineering can prevent one from breaking into the chain at any point, whenever one thinks human intuition and judgment should guide brute calculation.

It is essential, of course, for anyone to understand his instrument well enough to use it properly, but the computer is just like other commonplace instruments in this regard. Like any good tool, it should be used with respect. Applying "data reduction" techniques to voluminous data collected without adequate experimental design is a folly of the master not to be blamed on the servant. Computer folk have an acronym for it: GIGO, for "garbage in, garbage out."

X-ray crystallography is the most advanced of many instances in which similar instrumentation is being developed. Four experimental stations at the Cambridge Electron Accelerator, operated jointly by Harvard and M.I.T., are currently being connected to a time-shared computer at the Harvard Computing Center to provide a first link. A small computer at each experimental station converts instrument readings from analogue to digital form, arranges them in a suitable format and transmits them to the remote computer. There most data are stored for later detailed calculation; a few are examined to instruct each of the small machines to display information telling the experimenter whether or not his experiment is going well. Heretofore delays in conventional batch-processing procedures occasionally led to scrapping a long experiment that became worthless because poor adjustments could not be detected until all calculations were completed and returned.

This type of experiment is described as an "open loop" experiment, since the computer does not directly affect the setting of experimental controls. Closed-loop systems, where the experiment is directly controlled by computer, are currently being developed. Their prototypes can be seen in industrial control systems, where more routine, better-understood devices, ranging from elevators to oil refineries, are controlled automatically.

The problem of "reading" particle-track photographs efficiently has been a persistent concern of high-energy physicists. Here the raw data are not nearly as neat as they are in X-ray diffraction patterns, nor can photography as readily be bypassed. Automating the process of following tracks in bubble-chamber photographs to detect significant events presents very difficult and as yet unsolved problems of pattern recognition, but computers are now used at least to reduce some of the tedium of scanning the photographs. Similar forms of man-machine interaction occur also in the study of brain tumors by radioactive-isotope techniques. Where the problem of pattern recognition is simpler, as it is in certain types of chromosome analysis, there is already a greater degree of automation.

Let us now turn from the computer as instrument to the computer as actor, and to the subject of dynamic expression of theory. To understand clearly words such as "model," "simulation" and others that recur in this context, a digression is essential to distinguish the functional from the structural aspects of a model or a theory.

A robot is a functional model of man. It walks, it talks, but no one should be fooled into thinking that it is a man or that it explains man merely because it acts like him. The statements that "the brain is like a computer" or that "a network of nerve cells is like a network of computer gates, each either on or off," crudely express once popular structural theories, obviously at different levels. Both are now discredited, the first because no one has found structures in the brain that look anything like parts of any man-made computer or even function like them, the second because nerve-cell networks were found to be a good deal more complicated than computer networks.

A functional model is like the electrical engineer's proverbial "black box," where something goes in and something comes out, and what is inside is unknown or relevant only to the extent of somehow relating outputs to inputs. A structural model emphasizes the contents of the box. A curve describing the current passing through a semiconductor diode as a function of the voltage applied across its terminals is a functional model of this device that is exceedingly useful to electronic-circuit designers. Most often such curves are obtained by fitting a smooth line to actual currents and voltages measured for a number of devices. A corresponding structural model would account for the characteristic shape of the curve in terms that describe the transport of charge-carriers through semiconductors, the geometry of the contacts and so forth. A good structural model typically has greater predictive power than a functional one. In this case it would predict changes in the voltage-current characteristic when the geometry of the interfaces or the impurities in the semiconductors are varied.

If the black box is opened, inspiration, luck and empirical verification can turn a functional model into a structural one. Physics abounds with instances of this feat. The atom of Lucretius or John Dalton was purely functional. Modern atomic theory is structural, and the atom with its components is observable. The phlogiston theory, although functional enough up to a point, evaporated through lack of correspondence between its components and reality. Although the description of the behavior of matter by thermodynamics is primarily functional and its description by statistical mechanics is primarily structural, the consistency of these two approaches reinforces both.

The modern computer is a very versatile and convenient black box, ready to act out an enormous variety of functional or structural roles. In the physical sciences, where the script usually has been written in mathematics beforehand, the computer merely brings to life, through its program, a role implied by the mathematics. Isaac Newton sketched the script for celestial mechanics in the compact shorthand of differential equations. Urbain Leverrier and John Couch Adams laboriously fleshed out their parts in the script with lengthy and detailed calculations based on a wealth of astronomical observations. Johann Galle and James Challis pointed their telescopes where the calculations said they should and the planet Neptune was discovered. In modern jargon, Leverrier and Adams each ran Neptune simulations based on Newton's model, and belief in the model was strengthened by comparing simulation output with experiment. Computers now routinely play satellite and orbit at Houston, Huntsville and Cape Kennedy. Nevertheless, there is little danger of confusing Leverrier, Adams or a computer with any celestial object or its orbit. As we shall see, such confusion is more common with linguistic and psychological models.

The determination of protein structures provides an excellent example of how computers act out the implications of a theory. Finding a possible structure for a protein molecule covers only part of the road toward understanding. For example, the question arises of why a protein molecule, which is basically just a string of amino acid units, should fold into the tangled three-dimensional pattern observed by Kendrew. The basic physical hypothesis invoked for explanation is that the molecular string will, like water running downhill, fold to reach a lowest energy level. To act out the implications of this hypothesis, given an initial spatial configuration of a protein chain, one might think of calculating the interactions of all pairs of active structures in the chain, minimizing the energy corresponding to these interactions over all possible configurations and then displaying the resultant molecular picture. Unfortunately this cannot be done so easily, since no simple formula describing such interactions is available and, with present techniques, none could be written down and manipulated with any reasonable amount of labor. Sampling more or less cleverly the energies of a finite but very large number of configurations is the only possibility. An unsupervised computer searching through a set of samples for a minimum would, more likely than not, soon find itself blocked at some local minimum—unable, like a man in a hollow at the top of a mountain, to see deeper valleys beyond the ridges that surround him.

The close interaction of man and machine made possible by new "on line" time-sharing systems, graphical display techniques and more convenient programming languages enables Levinthal and his collaborators to use their intuition and theoretical insight to postulate promising trial configurations. It is then easy for the computer to complete the detail work of calculating energy levels for the trial configuration and seeking a minimum in its neighborhood. The human operator, from his intuitive vantage point, thus guides the machine over the hills and into the valley, each partner doing what he is best fitted for.

Even more exciting, once the details of the interactions are known theoretically, the X-ray diffraction pattern of the molecule can be calculated and compared with the original observations to remove whatever doubts about the structure are left by ambiguities encountered when going in the other direction. This closing of the circle verifies not only the calculation of molecular structure but also the theoretical edifice that provided the details of molecular interactions.

In this example the computer clearly mimics the molecule according to a script supplied by underlying physical and chemical theory. The computer represents the molecule with a sufficient degree of structural detail to make plausible a metaphorical identification of the computer with the molecule. The metaphor loses its force as we approach details of atomic structure, and the submodels that account for atomic behavior are in this case merely functional.

The remarkable immediacy and clarity of the confrontation of acted-out theory and experiment shown in the preceding example is by no means an isolated phenomenon. Similar techniques are emerging in chemistry, in hydrodynamics, and in other branches of science. It is noteworthy, as Don L. Bunker has pointed out, that computers used in this way, far from reducing the scientist to a passive bystander, reinforce the need for the creative human element in experimental science, if only because witless calculation is likely to be so voluminous as to be beyond the power of even the fastest computer. Human judgment and intuition must be injected at every stage to guide the computer in its search for a solution. Painstaking routine work will be less and less useful for making a scientific reputation, because such "horse work" can be reduced to a computer program. All that is left for the scientist to contribute is a creative imagination. In this sense scientists are subject to technological unemployment, just like anyone else.

In the "softer" emerging sciences such as psychology and linguistics the excitement and speculation about the future promise of the computer both as instrument and as actor tend to be even stronger than in the physical sciences, although solid accomplishments still are far fewer.

From the time modern computers were born the myth of the "giant brain" was fed by the obvious fact that they could calculate and also by active speculation about their ability to translate from one language into another, play chess, compose music, prove theorems and so on. That such activities were hitherto seen as peculiar to man and to no other species and certainly to no machine lent particular force to the myth. This myth (as expressed, for example, in *New Yorker* cartoons) is now deeply rooted as the popular image of the computer.

The myth rests in part on gross misinterpretation of the nature of a functional model. In the early 1950's, when speculation about whether or not computers can think was at the height of fashion, the British mathematician A. M. Turing proposed the following experiment as a test. Imagine an experimenter communicating by teletype with each of two rooms (or black boxes), one containing a man, the other a computer. If after exchanging an appropriate series of messages with each room the experimenter is unable to tell which holds the man and which the computer, the computer might be said to be thinking. Since the situation is symmetrical, one could equally well conclude that the man is computing. Whatever the decision, such an experiment demonstrates at most a more or less limited functional similarity between the two black boxes, because it is hardly designed to reveal structural details. With the realization that the analogy is only functional, this approach to the computer as a model, or emulator, of man loses both mystery and appeal; in its most naïve form it is pursued today only by a dwindling lunatic fringe, although it remains in the consciousness of the public.

In a more sophisticated vein attempts continue toward devising computer systems less dependent on detailed prior instructions and better able to approach problem-solving with something akin to human independence and intelligence. Whether or not such systems, if they are achieved, should have anything like the structure of a human brain is as relevant a question as whether or not flying machines should flap their wings like birds. This problem of artificial intelligence is the subject of speculative research described in the article by Marvin L. Minsky. Once the cloud of misapplied functional analogy is dispelled the real promise of using the computer as an animated structural model remains.

Mathematics has so far made relatively few inroads in either linguistics or psychology, although there are now some rather beautiful mathematical theories of language. The scope of these theories is generally limited to syntax (the description of the order and formal relations among words in a sentence). Based as they are on logic and algebra, rather than on the now more familiar calculus, these theories do not lend themselves readily to symbolic calculation of the form to which mathematicians and natural scientists have become accustomed. "Calculations" based on such theories must generally be done by computer. Indeed, in their early form some of these theories were expressed only as computer programs; others still are and may remain so. In such cases the language of programs is the language of science; the program is the original and only script, not just a translation from mathematics.

Early claims that computers could translate languages were vastly exaggerated; even today no finished translation can be produced by machine without human intervention, although machine-aided translation is technically possible. Considerable progress has been made, however, in using computers to manipulate languages, both vernaculars and programming languages. Grammars called phrase-structure grammars and transformational grammars supply the theoretical backdrop for this activity. These grammars describe sentences as they are generated from an initial symbol (say S for sentence) by applying rewrite rules followed (if the grammar is transformational) by applying transformation rules. For example, the rewrite rule $S \rightarrow SuPr$, where Su can be thought of as standing for subject and Pr as standing for predicate, yields the string $SuPr$ when it is applied to the initial symbol S. By adding the rules $Su \rightarrow John$ and $Pr \rightarrow sleeps$ one can turn this string into the sentence "John sleeps." Transformations can then be applied in order to turn, for example, the active sentence "John followed the girl" into the passive one "The girl was followed by John."

Under the direction of Susumu Kuno and myself a research group at Harvard has developed, over the past few years, techniques for inverting this generation process in order to go from a sentence as it occurs in a text to a description of its structure or, equivalently, to a description of how it might have been generated by the rules of the grammar. Consider the simple sentence "Time flies like an arrow." To find out which part of this sentence is the subject, which part the predicate and so on, a typical program first looks up each word in a dictionary. The entry for "flies" would show that this word might serve either as a plural noun denoting an annoying domestic insect or as verb denoting locomotion through the air by an agent represented by a subject in the third person singular.

The specific function of a word in a particular context can be found only by checking how the word relates to other words in the sentence, hence the serious problem of determining which of the many combinations of possible functions do in fact fit together as a legitimate sentence structure. This problem has been solved essentially by trying all possibilities and rejecting those that do not fit, although powerful tests suggested by theory and intuition can be applied to eliminate entire classes of possibilities at one fell swoop, thereby bringing the process within the realm of practicality.

A grammar that pretends to describe English at all accurately must yield a structure for "Time flies like an arrow" in which "time" is the subject of the verb "flies" and "like an arrow" is an adverbial phrase modifying the verb. "Time" can also serve attributively, however, as in "time bomb," and "flies" of course can serve as a noun. Together with "like" interpreted as a verb, this yields a structure that becomes obvious only if one thinks of a kind of flies called "time flies," which happen to like an arrow, perhaps as a meal. Moreover, "time" as an imperative verb with "flies" as a noun also yields a structure that makes sense as an order to someone to take out his stopwatch and time flies with great dispatch, or like an arrow.

A little thought suggests many minor modifications of the grammar sufficient to rule out such fantasies. Unfortunately too much is then lost. A point can be made that the structures are legitimate even if the sentences are meaningless. It is, after all, only an accident of nature, or for that matter merely of nomenclature, that there is no species of flies called "time flies." Worse yet, anything ruling out the nonexisting species of time flies will also rule out the identical but legitimate structure of "Fruit flies like a banana."

Still more confusing, the latter sentence itself is given an anomalous structure, namely that which is quite sensible for "Time flies. . ." but which is nonsensical here since we know quite well that fruit in general does not fly and that when it does, it flies like maple seeds, not like bananas.

A theory of syntax alone can help no further. Semantics, the all too nebulous notion of what a sentence means, must be invoked to choose among the three structures syntax accepts for "Time flies like an arrow." No techniques now known can deal effectively with semantic problems of this kind. Research in the field is continuing in the hope that some form of man-machine interaction can yield both practical results and further insight into the deepening mystery of natural language. We do not yet know how people understand language, and our machine procedures barely do child's work in an extraordinarily cumbersome way.

The outlook is brighter for man-made programming languages. Since these can be defined almost at will, it is generally possible to reduce ambiguity and to systematize semantics well enough for practical purposes, although numerous challenging theoretical problems remain. The computer is also growing in power as an instrument of routine language data processing. Concordances, now easily made by machine, supply scholars in the humanities and social sciences with tabular displays of the location and context of key words in both sacred and profane texts.

Psychologists have used programming languages to write scripts for a variety of structural models of human behavior. These are no more mysterious than scripts for the orbit of Neptune or the structure of hemoglobin. The psychological models differ from the physical ones only in their subject and their original language. Convincing empirical corroboration of the validity of these models is still lacking, and the field has suffered from exaggerated early claims and recurrent confusion between the functional and the structual aspects of theory. Psychology and the study of artificial intelligence are both concerned with intelligent behavior, but otherwise they are not necessarily related except to the extent that metaphors borrowed from one discipline may be stimulating to the other.

In actuality it is the languages, not the scripts, that are today the really valuable products of the attempts at computer modeling of human behavior. Several languages, notably John McCarthy's LISP, have proved invaluable as tools for general research on symbol manipulation. Research on natural-language data processing, theorem-proving, algebraic manipulation and graphical display draws heavily on such languages. Nevertheless, the computer as instrument is rapidly making a useful place for itself in the psychology laboratory. Bread-and-butter applications include the administration, monitoring and evaluation of tests of human or animal subjects in studies of perception and learning. . . .

It is also interesting to speculate on the use of on-line computers as tools for the investigation of the psychology of learning and problem-solving. Experiments in this area have been difficult, contrived and unrealistic. When the interactive computer serves as a problem-solving tool, it is also easily adapted to record information about problem-solving behavior. Here again the problem will not be the collection of data but rather devising appropriate experimental designs, since an hour's problem-solving session at a computer console can accumulate an enormous amount of data.

In short, computers are capable of profoundly affecting science by stretching human reason and intuition, much as telescopes or microscopes extend human vision. I suspect that the ultimate effects of this stretching will be as far-reaching as the effects of the invention of writing. Whether the product is truth or nonsense, however, will depend more on the user than on the tool.

The Social Sciences

ITHIEL DE SOLA POOL

Massachusetts Institute of Technology

The study of total societies is not novel. At least since Herodotus it has been a favorite pursuit of mankind. Historians and anthropologists, both of whom have engaged in this pursuit, have much to teach us about what can be done, how it should be done, and what cannot be done.

One of the first lessons of historiography is that while it is possible to study a total society, it is not possible to study a total society totally. There is no such thing as a complete description of even the smallest event. The job of the historian is to select. Out of the infinite complexity of reality he reports some few aspects which seem to him important, illustrative, indicative. Only a most naïve historian believes that these are "the facts." They are some facts, and a different historian with a different purpose and a different perspective might have chosen different facts.

Another key insight of both anthropologists and historians is that these myriad events are not independent or equally influential. Cultures have a certain unity. They are coherent. There are "pattens of culture." If one knows how a culture treats violence between siblings and between siblings and parents, one is likely to be able to guess how it treats violence among neighbors and citizens. A culture is not a random series of folkways and mores.

Out of these two insights—that one can never describe more than a small sample of the events that constitute a society and that these events are not all of equal significance, some being more indicative of fundamental patterns than others—grow many of the theses presented over the past century about how to study societies. However, these two propositions constitute

Reprinted from Ithiel de Sola Pool, "Computer Simulations of Total Societies," in Samuel Z. Klausner, ed., *The Study of Total Societies* (Garden City, New York: Doubleday & Company, Inc.—Anchor Books, 1967), pp. 45–58, 63–65, copyright © by the Bureau of Social Science Research. Used with the permission of Doubleday & Company, Inc.

an incomplete set of equations, an indeterminate system. They tell us something of great importance about how to describe a society, but they do not tell us enough to decide exactly what facts to include in a description of a total society. They do help us to distinguish profound from shallow work, good descriptions of societies from bad ones, for they provide some of the necessary criteria of criticism. But while such propositions do enable us to distinguish good work from bad work, they are inadequate to enable us to choose among good works.

There are an infinite number of alternative good, profound, sophisticated descriptions of the same society. There is no limit to the number of books that could be written about the United States with different selections of data to demostrate the values we have, the social structure, the prevailing beliefs and expectations, the traditions, the customs, etc.

An incomplete system is never satisfactory to a scholar. It is always an invitation to find the other variables and relationships not yet identified, which enable one to come up with a unique solution. For a century and a half, historians and social scientists have proposed more specific rules for the description of total societies. We have been told that we should start by looking at the mode of production and that we will find all other human relations to be reflections of them. We have been told to look at early child training and that the character of the culture is fixed by that. We have been told that one can enumerate sets of values and then categorize the society according to its selection among them. All of these insights have proved fruitful in that they have identified important factors, i.e., factors with greater predictive powers than some others we might choose. But there is no consensus on any such grand scheme for the description of a total society. . . .

From this consensus in discouragement arises an interest in simulation as a possible way of studying total societies. There is always a temptation to have exaggerated expectations for each new development in the social sciences. One such development which is indeed a major breakthrough for the analysis of complex systems is the technique of simulation. There is a natural inclination to feel that perhaps here at last we have a means for the study of total societies that will enable us to do what we have not so far been able to do, i.e., to find unique correct solutions. There is a tendency to believe that if only we can do a simulation of a society we will have achieved a complete, accurate representation about which we may say "this is the description of that society." There is some foundation for this view, although it is essentially incorrect.

A simulation is designed as an answer to the question of what would happen under some hypothetical future circumstances. One may simulate the effects of campaign strategies, or of marketing strategies, or of different organizational structures. In each case one is posing and trying to answer a "what if" question, to make a conditional prediction. Thus a simulation model must be sufficiently complex and the relationships within it sufficiently well defined to permit one to emerge with a specific prediction of the outcome under a specific hypothetical set of circumstances. (More accurately, if the model is stochastic, one must emerge with a specific prediction of the statistical distribution of a set of possible outcomes.) A simulation model therefore needs to be highly rigorous.

A simulation model may be contrasted with a typical *ceteris paribus* statement about variables taken pairwise, the kind of proposition of which most of social science consists. Typically if a social scientist says there is a tendency for Y to go one way if X goes a certain way, he makes it clear that he is not making a specific prediction of the value of Y because he is only singling out one of its determinants in a statement which is true only insofar as other things remain constant. The simulator, on the other hand, attempts to bring in all the variables that he considers to be important in determining the outcome, to state the relationships among them with such completeness that only one degree of freedom remains, and in that way to state what the actual outcome would be if a particular variable X were changed in a certain way.

That, however, does not mean that his model is the only valid model of the system that he is studying. A different model may be necessary to predict a different dependent variable in the same system. Indeed it is conceivable that two models would each predict the same dependent variable equally well. For example, to cite a familiar joke: one can measure the number of cows in a field either by counting the number of their heads or by counting the number of their legs and dividing by four. In the same way there may be many identically good and essentially substitutable indicators of a political or social fact. It may turn out that revolutions, for example, are equally well predicted by studying dissatisfaction of social classes or the accumulation of weapons in the hands of subversive organizations that are prepared to use them. If both conditions are necessary for a revolution and come about together then either one may provide an indicator in a simulation model as in any other social science procedure. Thus it is valid to say that a simulation of a total society, if successful, enables one to make specific predictions of outcomes but it is still not true that there is any one unique best simulation.

We have stated the essential requirement for a simulation to be that we can state the variables and their relations with sufficient precision to provide us with a determinate solution to a closed system. The system must provide a single definite prediction for each change that can be introduced in the independent variables. This degree of precision in outcome can be achieved in a number of ways, but one way or another there must be a decision rule that specifies what will happen to dependent variable Y when there is some specified hypothetical condition for each of the independent variables X_1, $X_2 ... X_0$.

Such decision rules can be of two kinds. They can either specify what the decision should be or they can specify who makes the decision. Thus a fundamental distinction among kinds of simulations is between mathematical or computer simulations on the one hand and human simulations (more often called games), on the other.

In the human simulation or game, a player is presented with a complex situation that has arisen out of the initial scenario and the previous moves of himself and other players. When it is his turn he makes a well-defined move;

he takes into account all the circumstances, decides what would or should be done by the person or institution whose role he is playing. He thus provides a completely definite outcome to the complex conditions hypothetically proposed. The decision rule simply specifies who chooses that outcome. Since the outcome is definite such games are properly called simulations. Since over several plays of the games different outcomes may be chosen by different players in identical circumstances, such games are properly described as stochastic simulations, and the result of the analysis of the simulation is more accurately described as a measure of the distribution of possible outcomes than as a single outcome.

In this paper we shall not consider human games further. We turn our attention to those simulations where the decision rules specify the outcome rather than specifying the authoritative decision maker. Such simulations are formal models, usually mathematical, and usually so complex that they can in practice only be operated with the aid of a computer. Their essential character, however, is neither that they are quantitative nor that they are computer-operated but that the formal model is sufficiently well defined to give a specific outcome. There must be no extra degrees of freedom producing indeterminacy of the outcome.

That may in many instances be achieved only by way of a device that to some degree resembles human simulation more than it resembles a typical mathematical model, namely a Monte Carlo device. One or more of the decision rules of the simulation may be to select a random number. Where that is done there is still a definite outcome because the random number gets selected. However, it will not provide the same outcome every time. Thus once again the analysis of such a simulation yields information only about the statistical distribution of those outcomes rather than about a single inevitable outcome. Nonetheless such simulations fit our definition. In each run there is a determinate prediction for any given hypothetical set of circumstances.

I cannot think of any complex formal simulation model which is completely non-quantitative. In principle, however, there can surely be one. Let us consider as a near example a simulation of English speech. Let us imagine that we have programed a computer to start with a word, then to look up in a dictionary what grammatical part of speech it is, then to provide another word that, according to English grammar, can follow such a part of speech, then to provide another word that could follow such a pair, etc., until one has had the computer create a grammatical English sentence. This is not an example of a completely non-quantitative simulation, given the many words in the dictionary that can follow any other given word or series of words. A Monte Carlo device would have picked one of them by means of a random number. Nonetheless, the role of quantity is clearly very minimal in such a simulation. So this simulation may serve as an example to show that computers are devices for proceeding according to rigorous formal rules. They are not necessarily arithmetic machines.

These points have a good deal to do with whether a simulation of a total society is feasible or not. On the one hand when people pooh-pooh this possibility by saying that there are many crucially important non-quantitative things to be said in describing a total society, they are raising an illegitimate objection. The statement is true but it may be an argument for the value of computer simulation rather than against. If the qualitative statements made are at all rigorous they can be said in computer language just as well as in English. There is nothing unambiguous that cannot be said in a computer interpretable code. Joseph Weizenbaum has said, and with some merit, that the time is not far off when no social science theory that cannot be reduced to a computer model will be given any credence or respect because its non-reducibility can only mean its ambiguity. The problem then is not one of quantity versus quality, it is one of precision versus ambiguity.

So while the simulation of total societies is possible, both in their qualitative as well as quantitative aspects, let us not assume it is always a sensible thing to do. It is sensible only if our thinking is sufficiently advanced so that we can replace all ambiguous expressions by definite ones. For numerous problems that is not the case. To cite an example from the field of marketing, no one has come up with a good simulation of advertising effectiveness, although many people have tried. The problem is that the phrase "advertising effectiveness" is ambiguous. Is it to be measured in awareness of the product, in its image, or perhaps, most obviously, in sales? The trouble with sales as a measure, however, is that they are also dependent upon many other factors besides advertising such as distribution, consumer habits, etc. Thus a simulation in which the dependent variable is sales must be a simulation not only of advertising but of the entire marketing process. In such a simulation we can give the concept of advertising effectiveness a definite meaning in terms of sales but at the same time we find that we have to introduce into a total market model several other equally ambiguous and poorly defined variables. Our state of knowledge about the total system is such as to make the attempt to produce a simulation of advertising effectiveness a futile one. It is possible to produce simulations of the frequency of advertising exposures, it is possible to produce simulations of the learning and forgetting process as the result of exposure; these are well-defined problems (partially qualitative, partially quantitative); they are, however, only sub-problems of the larger one of defining advertising effectiveness. The larger one is one we are not yet ready for.

Do not conclude from this that total society simulations are impossible. It is not always true that larger problems are harder to represent than smaller ones. That would be true only for models that are reductionistic. If, in general, models of larger systems had to be based on models of their subsystems then no social model would be possible except on the basis of a biological model and no biological model except on the basis of a chemical model. Clearly social scientists reject this view.

In short, there is no *a priori* reason why total society simulations should be impossible. They are not ruled out by the qualitative character of much of our understanding of social processes; they are not ruled out by the complexity of the reality they represent, nor by the number of subsystems they encompass.

On the other hand, the skeptic about total society simulation is probably more often right than wrong because more often than not we lack sufficiently definite information about the facts and relationships in the particular system that we have chosen to represent to provide an effective simulation. As a practical matter, the way to go about deciding whether a problem is ready for computer simulation is to try to sketch out with pencil and paper the relationships among the variables to see how far we think we can state them in a rigorous manner. A flow chart is a commonly used device for representing such a set of relationships, as is a series of simultaneous equations. It may also be done in verbal form.

But having a flow diagram does not make the problem computerizable. That depends on whether the relationships so easy to represent by lines and boxes can be stated rigorously. If one feels that one can state enough of them rigorously so that implications of a non-obvious character probably follow, then it probably is worthwhile undertaking the intellectual task of formalizing these relationships into a computer model. If the propositions form only a loose web and few non-obvious results would follow from their interactions, then what is called for is not computer simulation but more work on the underlying pieces of theory, for a simulation is merely an expression of theories. If the propositions are rich, numerous, and interrelated but so vaguely stated that nothing definite can be said to follow, then again what is called for is not computer simulation but more work on the conceptualization of the theory.

Let me cite some examples of situations that lent themselves to simulation. In 1960 I engaged in a simulation of presidential campaign strategies. The basic body of theory represented in that simulation was extremely simple as simulations go. There were some dozen variables in the equations. Three or four general propositions interrelated them in rather simple form. These general propositions are known in the social sciences as the theory of cross-pressure and are presented in the Berelson-Lazarsfeld-McPhee volume *Voting*. In addition to these few variables and simple propositions, however, there were in the model a very large number of parameters which had to be rather accurately estimated for the model to work. The propositions had to do with such matters as the probability of a person switching his vote as a function of party affiliation and issue attitudes. The parameters also measured such things as the number of people who had a certain issue attitude or who had a certain party affiliation in a particular state. In this instance the simulation was worth undertaking and worked well because it happened that there was a large body of good data for measuring these many parameters. That body of data was the accumulated public opinion polls from the previous decade. They provided us with good measures of such matters as civil rights attitudes or the degree of anti-Catholicism in different segments of the population.

This simulation may be described as a data-rich, theory-poor simulation. There was a small set of interrelated propositions that constituted the basis of the simulation. These taken by themselves were reasonably well established, obvious, and not very powerful. They provided a routine, however, for processing a large body of good data. Recognition of this potential justified attempting a computer simulation.

There are, however, other instances where one should proceed with a computer simulation although one has very little data or none at all. One might do so where the parameter estimates (so numerous in the election simulation) are relatively few and can be provided by a variety of guesses, and where the interest of the simulation focuses on a highly complex, well-defined structure of propositions. An example of such a simulation is the Crisiscom simulation in which we feed messages into the computer that represents two decision makers. In processing these messages, the computer follows rules representing a number of psychological propositions about attention, retention, and information handling. The computer attempts to select for attention and recall those messages which a human decision maker would retain under the specified circumstances. We use absolutely no empirical measures of the decision makers. We set values for forgetting, salience, and other parameters arbitrarily. The interest of the simulation is in the process, not in the measures of particular real world parameter values. If psychologists had not provided us in recent years with a number of well-defined models of cognitive processes, the attempt to develop such a computer simulation would be absurd.

For a simulation to be computerizable, there must somewhere exist a complex structure of propositions and/or data values. But note ·that we have had to use the expression and/or. It is not true that both must exist. Thus, it is not necessarily true that a simulation is no more useful than its data.

Let me illustrate with an example that is indeed a total society simulation. In recent years I have worked on this simulation in two different forms. In an earlier form, known as the Media-Mix simulation, it represents the flow of messages through the mass media to the American population. In a more sophisticated form on which I am currently working, the Comcom simulation,[1] it represents the flow of messages through the mass media to the populations of the Soviet Union and China. In each of these simulations a population of a few thousand individuals is represented in the computer. These represent a sample of the population of the country. Each individual has certain social and demographic characteristics and also certain media habits. In the simulation a flow of messages is released to them through various media; the computer calculates the probability of a message having been received by any given type of person after any given period of time.

This is a total society simulation. It is not a total simulation of a total society. It is a simulation of a single process, that of the flow of messages via the mass media, but for this selected aspect it examines, in a microcosm, the society as a whole. Any simulation of a total society is bound to be like that, i.e., it will be a representation of one or more aspects of society. Just as either visible light or infra-red can provide a picture of an object, though in each case actually only reporting the object through one particular index of it, so a society can be indexed by any one of a number of

[1] This project is supported by the Advanced Research Projects Agency under contract No. 920F-9717 with the Center for International Studies, M.I.T., and monitored by the Air Force Office of Scientific Research under grant number AF(49)638–1237.

pervasive aspects of it. Message flow is a good index because virtually every social process involves message flows. Exchange is another good index occurring in almost all social processes. So is power; so is role. That is why communication, economics, political science, and sociology are each effective ways of analyzing whole societies. Since these approaches are good ones for describing whole societies, they are good ones for simulating whole societies. In any case, the simulation we are here describing represents a society by its mass media message flow. . . .

We should concede that our simulation estimate is bound to be wrong. Like all measurements it has a standard error. Hypothetically, we may someday find out how big its standard error is. A statement of social science that is wrong but definite is better than the usual type whose main defense is a vagueness sufficient so that it cannot be *proved* wrong.

BIBLIOGRAPHY

American Behavioral Scientist. *Social Research with the Computer,* special issue, 1965, VIII, No. 9. (Contains descriptions of Comcom and Crisiscom simulations.)

Abelson, R. P. "Computer Simulation of Hot Cognition," in Tomkins, S. S., and Messick, S. *Computer Simulation of Personality,* New York: John Wiley, 1963.

————, and Bernstein, A. "A Computer Simulation Model of Community Referendum Controversies," *Public Opinion Quarterly,* 1963, XXVII, 93–122.

Archives Européennes de Sociologie. *Simulation in Sociology,* special issue, 1965, VI, No. 1.

Bauer, R. A., and Buzzell, R. D. "Mating Behavioral Science and Simulation," *Harvard Business Review,* 1964, XLII, 116–24.

Benson, O. "A Simple Diplomatic Game," in Rosenau, J. N. (ed.). *International Politics and Foreign Policy,* New York: The Free Press, 1961.

Berelson, Bernard, Lazarsfeld, Paul F., and McPhee, William N. *Voting,* Chicago: University of Chicago Press, 1954.

Beshers, J. M., and Reither, S. "Social Status and Social Change," *Behavioral Science,* 1963, VIII, 1–14.

Borko, H. (ed.). *Computer Applications in the Behavioral Sciences,* Englewood Cliffs, N.J.: Prentice Hall, 1962.

Browning, R. P. "Computer Programs as Theories of Political Processes," *Journal of Politics,* 1962, XXIV, 562–82.

Clarkson, G. P. S., and Simon, H. A. "Simulation of Group Behavior," *American Economic Review,* 1960, IV, 920–31.

Coleman, J. S. "Mathematical Models and Computer Simulation," in Faris, R. E. L. *Handbook of Modern Sociology,* Chicago: Rand McNally, 1964.

————. "The Use of Electronic Computers in the Study of Social Organizations," *Archives Européennes de Sociologie,* 1965, VI, 89–107.

Guetzkow, H. (ed.) *Simulation in Social Science,* Englewood Cliffs, N.J.: Prentice Hall, 1962.

Gullahorn, J., and Gullahorn, J. "A Computer Model of Elementary Social Behavior," in Feigenbaum, E. and Feldman, J., *Computers and Thought,* New York: Mc-Graw Hill, 1965.

Hagerstrand, T. "A Monte Carlo Approach to Diffusion," in *Archives Européennes de Sociologie,* 1965, VI, 43–67.

Lazarsfeld, Paul F. *Radio and the Printed Page,* New York: Duell, Sloan and Pearce, 1940.

McPhee, W. N. *Formal Theories of Mass Behavior,* New York: The Free Press, 1963.

Pool, I. de S., Abelson, R. P., and Popkin, S. L. *Candidates, Issues, and Strategies: A Computer Simulation of the 1960 and 1964 Elections,* Cambridge, Mass.: M.I.T. Press, 1965.

————, and Kessler, A. "The Kaiser, the Tsar, and the Computer: Information Processing in a Crisis," *American Behavioral Scientist,* 1965, VIII, 31–39.

Popkin, S. L. "A Model of a Communication System," *American Behavioral Scientist,* 1965, VIII, 8–12.

Rainio, K. "Social Interaction as a Stochastic Learning Process," *Archives Européennes de Sociologie,* 1965, VI, 68–88.

————. *A Stochastic Theory of Social Contacts,* Copenhagen: Munksgaard, 1962.

Tomkins, S. S., and Messick, S. *Computer Simulation of Personality,* New York: John Wiley, 1963.

The Arts

JOHN R. PIERCE

Bell Telephone Laboratories

Two reproductions of prints by Harunobu hang on the right wall of my office. I know what I think of these. On the left I have reproductions of paintings by Ingres and David. I know what I think of these, too. When I look at the wall opposite my desk, I am a little puzzled. There I see a buff painting, five feet long and ten and a half inches wide. I understand the inscription in the lower left; it reads *Pour John Pierce, amicalement, Jean Tinguely, Avril 1962.*

The painting itself consists of strokes of red, turquoise, and gray ink, generally to the right and downward. Most of the strokes are accented at the beginning. The pattern of strokes is densest and widest a little above the middle, and the turquoise and gray strokes are nearly vertical toward the bottom of the picture. The general effect is Japanese.

This painting is the product of a stupid machine of clanking metal parts, a machine devised and built by the talented constructor of the jiggling "metametics" which have been shown in many countries, and of the celebrated "self-destroying machine" which partially succeeded some years ago in the courtyard of the Museum of Modern Art. Tinguely once built many painting machines similar to the one that created my picture, and sold them to a variety of people, including Nelson Rockefeller.

If I didn't like the painting on my wall, I wouldn't have it there. I am astonished that in some sense it is the product of a machine. But I am appalled when I think that a few hundred feet to my left there resides a machine, an electronic computer, which is to Tinguely's machine as Newton is to an earthworm. What sort of art can we expect from a comparative genius of a machine when a clanking metal monstrosity can produce a picture of at least dubious merit?

While intellectual visionaries have busied themselves asserting that the computer will outstrip man in his intellectual endeavors, and will manage wisely where the executive now mismanages, a less noisy few have approached the computer with artistic intentions, hoping to elicit from it something more patterned and of greater impact than chaos.

Indeed, a similar quest goes back well beyond the digital computer. Many years ago, Marcel Duchamp, who painted *Nude Descending a Staircase,* allowed one-meter-long white threads to fall from a height of one meter upon a flat surface. Some were framed and I have seen them. In the curved order imposed by the stiffness of the thread and the random configuration resulting from its passage through the air, there is a mixture of the graceful and the unexpected. Too, by adding a repeated symmetry to a random pattern of bits of colored glass, the kaleidoscope has pleased many generations of children and adults.

From the remote past to the very present, human beings have *incorporated* geometrical forms and psychological tricks in their art. The straight lines and rectangles of Mondrian have a geometrical regularity which we might associate with a machine, and the subtle curves of Op Art remind us both of mathematical curves and of the psychological texts on perception and optical illusions from which they are drawn. When the artist approaches science and the machine, will the machine perhaps approach the complexity and surprise which we associate with the human artist?

I don't know who first used the electronic computer to produce patterns of some originality and interest, but it may have been Dr. Bela Julesz of the Bell Telephone Laboratories. In studying properties of vision, Dr. Julesz caused the computer to generate patterns of black and white dots within squares, in which just a little order was imposed upon randomness. The result was so pleasing that a Japanese publisher used the pattern on the cover of a translation of the book in which it appeared.

Others have invoked the computer as an artist with more direct motives. Thus, the computer has been used not only to solve the equations of motion of a particular kind of satellite in orbit around the earth, but also to create an animated motion picture showing the satellite at first tumbling around in its orbit and finally aligning itself radially so that it points at the earth. Another programmer has caused the computer to produce a whole animated instructional motion picture showing rolling balls, the operation of the computer itself, and titles that rise across the screen, expand, and dissolve. The result is far from Walt Disney in skill, but much cheaper in cost. And one ingenious programmer did manage to make the computer draw pictures of Mickey Mouse's head as seen from any chosen direction.

This is serious work. Scientists and engineers want to present data in graphical and even in moving form, and they want to see what proposed devices and structures will look like from various angles. In some cases, the computer can produce the required drawings, or sequences of drawings, much more quickly and cheaply than could the most skilled draftsman.

But people have been tempted beyond these practical essays in computer art. In fact, one ingenious man, A. M. Noll, caused the computer to generate drawings in the style of Piet Mondrian, drawings consisting of short, heavy vertical and horizontal lines rather randomly arranged on a sheet of paper. Then Noll carried out a psychological experiment. He showed 100 people an original Mondrian drawing and a drawing made by the computer in the style of Mondrian. He asked them to decide which drawing was artistically better, and which was produced by a machine. Of all those asked, only 28 percent correctly identified the computer picture, and 59 percent preferred it to the Mondrian. However, people who said they disliked or were indifferent to modern art were equally divided in preferring the computer picture or the Mondrian; but people who said they liked modern art preferred the computer picture three to one. I don't know whether this is overestimating the computer's artistic ability or underestimating Mondrian's.

Noll has taken the computer far beyond imitation. In his use of the computer, he always prescribes some order but leaves the drawing partly to chance. By these means he has produced a weaving pattern formed by a self-intersecting line, patterns of lines splattered over a page, and even pairs of drawings which, when viewed through a stereoscope, give the effect of many lines hanging in space, much like the *Orpheus and Apollo* of Richard Lippold in the foyer of the Philharmonic Hall in Lincoln Center in New York City, but without any supporting wires at all. I feel driven to the fatuous comment: It's fascinating, but is it *art?*

Whether the computer, man, or both together create the art of the future, it is likely that man rather than the computer will enjoy it, and the place of a good deal of art is in the home. Today we have books and magazines, TV, slides, and primitive forms of 3-D. But the future holds something better in store for us. Emmet Leith and his colleagues at the University of Michigan have produced a visual effect as real as looking through a window.

By illuminating an object to be "photographed" with a coherent beam of light from a laser, that much-vaunted marvel of quantum electronics, Leith produces what is called a hologram, a wavy pattern of ultrafine lines on a photographic plate. When this hologram is illuminated by a laser, a person looking through it sees behind the hologram what appears to be a very solid three-dimensional version of the object that was used in producing the hologram. The whole object is represented in all parts of the hologram. When one moves his head, it is just as if he were looking through a window. If a less interesting detail is in front of an interesting part of the object, one merely has to move his head to see around it. Imagine such solidity, such rotundity, which goes far beyond that of 3-D movies or the old-fashioned stereoscope. At present, one can achieve this effect in only one color, and in still pictures, but who knows what the future will bring?

A computer is blind, deaf, and dumb, and it produces visual art only because someone forces it to. A computer can just as well produce a numerical description of a sound wave—in fact, a description of any sound wave. Don't think that people who are ear-minded rather than eye-minded have neglected to make computers produce sound. In the earliest attempts, a computer was made to play simple tunes in buzzes or squeaks, but we are now far beyond that point. A Decca record of 1962, *Music from Mathematics,* shows that the computer can play tunes in a variety of tone qualities, imitating plucked strings, reed instruments and other common effects, and going beyond these to produce shushes, garbles, and clunks that are unknown in conventional music. Further, the computer can even speak and sing. In the record I refer to, the computer actually sings *A Bicycle Built for Two*—to its own accompaniment.

Today, scientists and musicians at Massachusetts Institute of Technology, Bell Telephone Laboratories, Princeton University, and the Argonne National Laboratory are trying to make the computer play and sing more surprisingly and more mellifluously. As a musical instrument, the computer has unlimited potentialities for uttering sound. It can, in fact, produce strings of numbers representing any conceivable or hearable sound. But as yet, the programmers are somewhat in the position of a savage confronted with the grand piano. Wonderful things could come out of that box if only we knew how to evoke them.

While some mathematical musicians, and musical mathematicians, are trying to use a computer as a super orchestra, others are following a much older line of endeavor, which goes back to Mozart. Mozart provided posterity with a collection of assorted numbered bars in three-eight time together with a set of rules. By throwing dice to obtain a sequence of random numbers, and by using these numbers in choosing successive bars according to simple rules, even the nonmusical amateur can "compose" an almost endless number of little waltzes, which sound something like disorganized Mozart. Joseph Haydn, Maximilian Stadler, and Carl Philipp Emanuel Bach are said to have produced similar random music.

In more recent times, the inimitable John Cage has used a random process in the selection of notes. Indeed, there is a whole school who believe that chance is better than judgment and that a composition would be freshest if the composer guided it in a general way only, letting the individual notes fall where they will.

Some of the early experiments in this direction were as primitive as shaking a pen at a sheet of music paper and adding stems to the ink dots. Since the coming of the computer, chaos has entered music more scientifically. In 1956, the Burroughs Corporation announced it had used the computer to generate music, and in 1957 it was announced that Dr. Martin Kline and Dr. Douglas Bolitho had used the Datatron to write popular melodies. Jack Owens set words to one—which was played over the ABC network as Pushbutton Bertha. In 1957, F. P. Brooks, Jr., A. L. Hopkins, Jr., P. G. Neumann, and W. V. Wright published an account of the statistical composition of music on the basis of extensive statistical data on hymn tunes.

Perhaps the most ambitious early attempt was that of Lejaren A. Hiller, Jr., and Leonard M. Isaacson of the University of Illinois, who succeeded in formulating the rules of four-part first-species counterpoint in such a way that a computer could choose notes randomly and reject them if they violated these rules. Music so generated, together with other partially random, partially controlled music, was published in 1957 as the *Illiac Suite for the String Quartet.*

Since that time, the computer has come to function in a dual capacity, as an orchestra playing its own compositions. J.C. Tenney, who is now in the music department at Yale, has been a strong advocate of this approach. As a composer, he provides general guidance to the computer as to high or low, slow or fast, loud or soft, and some guidance as to timbre. Within specified ranges that change with time, the computer chooses the notes at random and plays them according to its own directions. The results are surprising in many ways. However unpredictable chance may be, it has a sort of uniformity that seems to preclude the kind of surprise one finds in Haydn's *Surprise Symphony,* that is, a carefully calculated loud effect following a soft passage. Perhaps the composer should provide the computer with more or less guidance, or perhaps guidance should be built into the computer.

Musicians of the modern school condemn, or at least wish to depart from, traditional musical devices and forms, but this hasn't kept musical scholars from analyzing music to see just what the form consists of. Harry F. Olson and his co-workers at RCA have already put Stephen Foster's melodies through the wringer and caused a computer to generate Foster-like tunes. In principle, what makes Mozart like Mozart, Haydn like Haydn, Wagner like Wagner is not beyond analysis. I would be very surprised if someone could cause a computer to produce good and original Mozart at the push of a button. I wouldn't be surprised at someone's making the computer sound something like Haydn or Mozart or Bach.

As in the case of the visual arts, new science and technology have much to offer in the reproduction as well as in the creation of the sounds of the future. It's a commonplace that listening to a stereo system, however good, isn't like hearing an orchestra in a concert hall. Yet it is not beyond the ability of science to create in one very particular place in a room the exact environment of sound that one would experience in a concert hall. Manfred Schroeder at the Bell Telephone Laboratories has shown how this can be done. He uses the computer to process the sounds that will be played over a pair of loudspeakers, so that in the vicinity of a person's head he creates the exact acoustical environment of a huge reverberant hall. This effect is uncanny. It is much fuller than a stereo system, and it is very different from hearing something through headphones.

The ability to localize sounds outside one's head, the feeling of being immersed in sound, depends on the way in which what one hears changes as he moves his head slightly. Schroeder cleverly simulates the sound near the head so that when one turns his head slightly, this has just the effect on the sounds he hears that it would if he were in an auditorium. At present this has to be carried out at great cost in an anechoic or echoless chamber, a large and expensive room with sound-absorbing walls. But who is to say that at some not-distant date it may not be possible to create exactly the same effect in any easy chair at home?

The question of whether a computer can be made to *write* as well as to draw, compose, and play is no less provocative. The manufacture of *meaningful* prose and poetry, as of art and music, is a challenge that may or may not be beyond the capacity of the computer, but the composition of striking new words and sentences is certainly well within the realm of mechanization. As part of a linguistic experiment conducted at the Bell Telephone Laboratories in 1961, for example, Dr. Melvin Hinich caused a computer to generate a number of rather compelling sentences which, considered as a single composition, might be said to substantiate my belief that the artistic utterances of mechanical chance and of contemporary avant-garde writers are approaching each other so closely as to come into competition. Wrote the computer.

> this is shooting
> this seems to be sleeping
> a vapid ruby with a nutty fan lies seldom below this tipsy noise.
>   this cute snake by that wet pig is clawing coolly to a weak pig
> any black otter below a holy fan is poking hotly in that furry ape to killing from this tipsy bat
> a fake mud on this cute hero is seldom sipping that bad moose below a tipsy house in moving from this tipsy creep.

Rather vivid imagery, I think, if a bit less than illuminating. But one doesn't need a computer, or even a beat poet, to generate such literary gems. One can do it with a pencil and paper and dice, or even with a group of cooperative human beings. C. E. Shannon, the inventor of information theory, demonstrated this many years ago when he chose letters on the basis of the probability that they would follow preceding letters. This led to the creation of some new words: *deamy, ilonasive, grocid, pondenome.* To me *deamy* has a pleasant sound. If someone said I had a deamy idea, I would take it in a complimentary sense. On the other hand, I'd hate to be denounced as ilonasive. I would not like to be called grocid, perhaps because it reminds me of gross, groceries, and gravid. Pondenome is at least dignified.

Shannon carried this further, and chose words on the basis of their probability of following other words. Anyone can carry on a similar process easily, as a sort of parlor game. You can write, say, three grammatically connected words in a column at the top of a slip of paper. You can then show these to a friend and ask him to make up a sentence in which the three words occur and to add the next word of this sentence. You then fold over the top word of the four, show the remaining three to yet another friend and get an additional word from him, and so on. After I had canvassed 20 friends, I had the following: "When morning broke after an orgy of wild abandon, he said her head shook vertically aligned in a sequence of words signifying what."

One can invent more complicated means for producing grammatical sentences that wander over the same ground but never exactly repeat. By using a chart of phrases and flipping heads or tails, I obtained the following interesting item: "The Communist Party investigated the Congress. The Communist Party purged the Congress and destroyed the Communist Party and found evidence of the Congress." This could go on forever, always grammatical and never exactly repeating, but I don't know to what end.

There have been other experiments with random language, of various sophistication and success. In 1946, a Yale undergraduate walked into the Sterling Memorial Library at Yale, picked a direction at random, took a book off a shelf at random, selected a page and a sentence on that page at random, and repeated the procedure until he had produced a 20-line "poem." This was accepted for publication as a legitimate man-made composition by the *Yale Poetry Review,* but the young man got cold feet at the last moment and withdrew the manuscript.

Though the Yale poem is long since lost, I can regale you with the following poem of my own, which I "composed" in about ten minutes by gleaning random quotes from a book selected at random from my shelves: *Great Science Fiction by Scientists,* edited by Groff Conklin.

> The Dictator shoved his plate aside with a petulant gesture
> The homely smile did not dismay him
> He was still not quite sure what had happened
> "I doubt if they starved," said Pop quietly
> The needle was near the first red mark
> Well, I merely pose the question.

Author William Burroughs is less painstaking and squeamish than was the student at Yale. He writes his books by cutting up already-written material and pasting the pieces together after mixing. With this montage technique he has written five or six books; the best known is *Naked Lunch.* Recently, I read in the press that a young student had succeeded in producing quite effective modern poems by a process that involved choosing lines or phrases entirely at random. I found the effect striking, but I am too old-fashioned to prefer it to Milton.

A group up at MIT some years ago tried another tack. They asked the computer to plot a simple story, choosing at random, for instance, whether the shot fired by the sheriff killed the bad man, or vice versa. To my mind, Zane Grey did better; but then, this was a very early MIT effort.

Matters of art aside, there is no question that machines other than computers, and computers themselves, have made pictures, have played music, have made music, and have constructed a semblance of English. What I am to think of this I find as hard to know as what I am to think of Jean Tinguely's painting that hangs on my wall. Some of what has come out of the computer isn't as bad as the worst of man-made art, but it certainly isn't as good as the best. The computer is a great challenge to the artist. It enables him to create within any set of rules and any discipline he cares to communicate to the computer. Or, if he abandons discipline, he may leave everything to chance and produce highly artistic noise.

I am sure that time will extend all the possibilities and opportunities for artistic creation and reproduction that I have described, and will bring them economically within the reach of the general public. Come tomorrow, we will be able to close our eyes and hear in our living room something completely indistinguishable from what we might hear in a concert hall or a theater. And it may be that we will also be able to open our eyes and *see,* in all its solidity, what we might see in the concert hall or the theater. What will we see? What will we hear? We may hear a poem written by a computer, sung in a computer voice, to an accompaniment of computer-generated and computer-played music. Perhaps we will see a ballet of computer-generated figures dancing in computer-generated patterns.

Scientists can only provide the means for doing this. Artists must school the computer if this is to become reality. I think that it isn't too early for artists and programmers to study man and his arts on the one hand, and the computer and its potentialities on the other, hotly and realistically. We must decide whether men and machines should work together gravely or wackily to produce works that are portentous or delicious. The choice is open, and I hope it won't be made too solemnly.

Libraries and Information

J. C. R. LICKLIDER
Massachusetts Institute of Technology

We need to substitute for the book a device that will make it easy to transmit information without transporting material, and that will not only present information to people but also process it for them, following procedures they specify, apply, monitor, and, if necessary, revise and reapply. To provide those services, a meld of library and computer is evidently required. . . . In thinking about procognitive systems—systems to facilitate man's interaction with transformable information—we should be prepared to reject the schema of the physical library—the arrangement of shelves, card indexes, check-out desks, reading rooms, and so forth. That schema is essentially a response to books and to their proliferation. If it were not for books, and for the physical characteristics of books. . .there would be no *raison d'être* for many parts of the schema of the physical library. We should be prepared to reject the schema of the physical book itself, the passive repository for printed information. That involves rejecting the printed page as a long-term storage device, though not for short-term storage and display. . . . What is of value for our purpose is. . .the schema in which a man sits at a desk, writes or draws on a surface with a stylus, and thereby communicates to a programmed information processor with a large memory. It is the mental image of the immediate response, visible on the oscilloscope, through which the computer acknowledges the command and reports the consequences of carrying it out—in which the computer acknowledges the question and presents an answer. Without such schemata in mind, one cannot think effectively about future systems for interaction with the body of knowledge. With such schemata, and enough others suggested by experiences in other contributory fields, perhaps conceptual progress can be made. . . . Extrapolation, however uncertain, suggests that the basic "mechanical" constraints will disappear: Although the size of the body of knowledge, in linear measure of printed text, is almost astronomical (about 100,000,000 miles), although that measure is increasing exponentially, and although the technology that promises to be most helpful to us in mastering knowledge is still young and weak, time strongly favors the technology. The technology, too, is growing exponentially and its growth factor is perhaps 10 times as great as the growth factor of the corpus. Moreover, the technology is not yet near any fundamental physical limits to development. Thus in the present century, we may be technically capable of processing the entire body of knowledge in almost any way we can describe; possibly in ten years and probably within twenty, we shall be able to command machines to "mull over" separate subfields of the corpus and organize them for our use—if we can define precisely what "mulling" should mean and specify the kind of organization we require. . . .

A man, reading eight hours a day every work day, at a speed appropriate for novels, could just keep up with new "solid" contributions to a subfield of science or technology. It no longer seems likely that we can organize or distill or exploit the corpus by passing large parts of it through human brains. It is both our hypothesis and our conviction that people can handle the major part of their interaction with the fund of knowledge better by controlling and monitoring the processing of information than by handling all the detail directly themselves. . . . [Therefore] a basic part of the overall aim for procognitive systems is to get the user of the fund of knowledge into something more nearly like an executive's or commander's position. He will still read and think and, hopefully, have insights and make discoveries, but he will not have to do all the searching himself nor all transforming, nor all the testing for matching or compatibility that is involved in creative use of knowledge. He will say what operations he wants performed upon what parts of the body of knowledge, he will see whether the result makes sense, and then he will decide what to have done next. Some of his work will involve simultaneous interaction with colleagues and with the fund of stored knowledge. Nothing he does and nothing they do will impair the usefulness of the fund to others. Hopefully, much that one user does in his interaction with the fund will make it more valuable to others. . . .

Economic criteria tend to be dominant in our society. The economic value of information and knowledge is increasing. By the year 2000, information and knowledge may be as important as mobility. We are assuming that the average man of that year may make a capital investment in an "intermedium" or "console"—his intellectual Ford or Cadillac—comparable to the investment he makes now in an automobile, or that he will rent one from a public utility that handles information processing as Consolidated Edison handles electric power. In business, government, and education, the concept of "desk" may have changed from passive to active: a desk may be primarily a display-and-control station in a telecommunication-telecomputation system—and its most vital part may be the cable ("umbilical cord") that connects it, via a wall socket, into the procognitive utility net.

Thus our economic assumption is that interaction with information and knowledge will constitute 10 or 20 per cent of the total effort of the society, and the rational economic (or socioeconomic) criterion is that the society be more productive or more effective with procognitive systems than without. . . .

To some extent, of course, the severity of the criteria that procognitive systems will be forced to meet will depend upon whether the pro- or anti-intellectual forces in our society prevail. It seems unlikely that widespread support for the development of procognitive systems will stem from appreciation of "the challenge to mankind," however powerful that appreciation may be in support of space efforts. The facts that information-processing systems lack the sex-symbolizing and attention-compelling attributes of rockets, that information is abstract whereas the planets and stars are concrete, and that procognitive systems may be misinterpreted as rivaling man instead of helping him—these facts may engender indifference or even hostility instead of support. . . .

The foregoing considerations suggest that the economic criterion will be rigidly enforced, that procognitive systems will have to prove their value in dollars before they will find widespread demand. If so, procognitive systems will come into being gradually, first in the richest, densest areas of application, which will be found mainly in government and business, and only later in areas in which the store of information is poor or dilute. Close interaction with the general fund of knowledge, which is on the whole neither rich nor dense, will be deferred, if these assumptions are correct, until developments paid for by special procognitive applications have made the broader effort practicable. Such a "coattail" ride on a piecemeal carrier may not be the best approach for the nation or the society as a whole, but it seems to be the most probable one. . . .

A HYPOTHETICAL EXAMPLE OF USE OF THE PROCOGNITIVE SYSTEM

Perhaps the best way to consolidate the picture that we have been projecting, one part at a time, is to describe a series of interactions, between the system and a user who is working on a substantive problem that requires access to, and manipulation of, the fund of knowledge. . . .

Friday afternoon—I am becoming interested, let us say, in the prospect that digital computers can be programmed in such a way as to "understand" passages of natural language. (That is a 1964 problem, but let us imagine that I have available in 1964 the procognitive system of 1994.) . . .

Immediately before me on my console is a typewriter that is, in its essentials, quite like a 1964 office typewriter except that there is no direct connection between the keyboard and the marking unit. When I press a key, a code goes into the system, and the system then sends back a code (which may or may not be the one I sent), and the system's code activates the marking unit. To the right of typewriter, and so disposed that I can get into position to write on it comfortably if I rotate my chair a bit, is an input-output screen, a flat surface 11″ × 14″ on which the system and I can print, write, and draw to each other. . . .

In a penholder beside the screen is a pen that can mark on the screen very much as an ordinary pen marks on paper, except that there is an "erase" mode. The coordinates of each point of each line marked on the screen are sensed by the system. The system then "recognizes" and interprets the marks. Inside the console is a camera-projector focused upon the screen. Above the chair is a microphone. The system has a fair ability to recognize speech sounds, and it has a working vocabulary that contains many convenient control words. Unfortunately, however, my microphone is out of order. There is a power switch, a microphone switch, a camera button, and a projector button. That is all. The console is not one of the high-status models with several viewing screens, a page printer, and spoken output.

The power is on, but I have not yet been in interaction with the system. I therefore press a typewriter key—any key—to let the system know the station is going into operation. The system types back, and simultaneously displays upon the screen:

14:23 13 November 1964

Are you J. C. R. Licklider?

(The system knows that I am the most frequent, but not the only, user of this console.) I type "y" for yes, and the system, to provide a neat record on the right-hand side of the page, types:

J. C. R. Licklider

and makes a carriage return. (When the system types, the typewriter operates very rapidly; it typed my name in a fifth of a second.) The display on the screen has now lost the "Are you . . ." and shows only the date and name. Incidentally, the typing that originates with me always appears in red; what originates in the computer always appears in black.

At this early stage of the proceedings, I am interacting with the local center, but the local center is also a subsystem of systems other than the procognitive system. Since I wish to use the procognitive system, I type

Procog

and receive the reply:

You are now in the Procognitive System.

To open the negotiation, I ask the procognitive system:

What are your descriptor expressions for:
    computer processing of natural language
    computer processing of English
    natural-language control of computers
    natural-language programming of computers
DIGRESS

At the point at which I wrote "DIGRESS," it occurred to me that I might in a short while be using some of the phrases repeatedly, and that it would be convenient to define temporary abbreviations. The typed strings were appearing on the display screen as well as the paper. (I usually leave the console in the mode in which the information, when it will fit and not cause delay, is presented on both the typewriter and the screen.) I therefore type:

define temp

On recognizing the phrase, which is a frequently used control phrase, the system locks my keyboard and takes the initiative with:

define temporarily

via typewriter? via screen?

I answer by swiveling to the screen, picking up the pen, and pointing to screen on the screen. I then point to the beginning and end of computer processing, then to the c and the p, and then to a little square on the screen labeled "end symbol." (Several such squares, intended to facilitate control by pointing, appear on the screen in each mode.)

In making the series of designations by pointing just described, I took advantage of my knowledge of a convenient format that is available in the mode now active. The first two pointings designate the beginning and end of the term to be defined, and the next pointings, up to "end symbol," spell out the abbreviation. (Other formats are available in the current mode, and still others in other modes.) If my microphone had been working, I should have said "Define *cee pee* abbreviation *this*" and drawn a line across computer processing as I said *"this."* The system would then have displayed on the screen its interpretation of the instruction, and then (after waiting a moment for me to intervene) implemented it.

Next, I define abbreviations for "natural language" (nl), "computer" (comp), and "programming" (prog). (Unless instructed otherwise, the system uses the same abbreviation for singular and plural forms and for hyphenated and unhyphenated forms.) And finally, insofar as this digression is concerned, I touch a small square on the screen labeled "end digression," return to the typewriter, and type:

comp understanding of nl
comp comprehension of semantic relations?§

The question mark terminates the query, and the symbol § tells the system not to wait for further input from me now.

Because the system's over-all thesaurus is very large, and since I did not specify any particular field or subfield of knowledge, I have to wait while the requested information is derived from tertiary memory. That takes about 10 seconds. In the interim, the system suggests that I indicate what I want left on the display screen. I draw a closed line around the date, my name, and the query. The line and everything outside it disappear. Shortly there-after, the system tells me:

Response too extensive to fit on screen. Do you wish short version, multipage display, or typewriter-only display?

Being in a better position to type than to point, I type:

s

That is enough to identify my preference. The short version appears on the screen but not the typewriter—(I rather expected it there too) :

Descriptor expressions:
   1. (natural language) $\wedge$ (computer processing of)
   2. (natural language) $\wedge$ (on-line man-computer interaction)
   3. (natural language) $\wedge$ (machine translation of)
   4. (natural language) $\wedge$ (computer programming)
   5. (computer program) $\wedge$ (semantic net)
   6. (compiler) $\wedge$ (semantic)
Descriptor inclusions:
   7. (natural language) includes (English)
Phrase equivalences:
   8. (control of computers) $\simeq$ (on-line man-machine interaction)
   9. (programming of computers) $\simeq$ (computer programming)
  10. (semantic relations) $\simeq$ (semantic nets)

                                              [END]

I am happy with 1 and especially happy with 2. I am curious about 5. I think I should look into 4. Number 3 frightens me a little, because I sus-pect it will lead to too much semirelevant information. I think I recognize 6 as a false lead into the field of program compiling, in which field, although "syntax" seems to be used in essentially the linguist's sense, "semantics" is used in what appears to me to be an inscrutable kind of jargon. Accordingly, I type to the system:

Prepare bibliographies with abstracts (1, 2, 4, 5).
Best 25 references each. When will they be available?

Before turning control over to the system, I decide to ask a substantive question, even though I realize I do not yet know enough about the subject to inquire intelligently. Using the control word, "QUESTION," to establish the mode, I ask:

QUESTION: Has any digital-computer system responded to questions phrased in ordinary English, or other natural language, by giving correct answers in natural language? §

The system replies first to "When will they be available?"

Requested bibliographies will be available at about 18:00.
Do you want them delivered, or will you call, or will you request typeout?

Unfortunately, my office is not located near a pneumatic-tube station. I rely heavily on typeouts made by the system when I am away from the console. I reply:

Type them out as soon as they are available.

The system reminds me:

Please leave your console power on!

Then it replies to the "QUESTION":

Cannot give yes-no answer to QUESTION. Approaches to "yes" are described in the following:
1. B. F. Green, A. K. Wolf, C. Chomsky, and K. Laughery, Baseball: An Automatic Question-Answerer, Proc. Western Joint Computer Conf., 19, 219–224, 1961.
2. S. Klein and R. F. Simmons, A Computational Approach to Grammatical Coding of English Words, J. Assoc. Computing Machinery, 10, 334–347, 1963.
3. S. Klein and R. F. Simmons, Syntactic Dependence and the Computer Generation of Coherent Discourse, Mechanical Translation (entering system).

The foregoing must suffice to suggest the nature of the interaction at the level of key pressing and pointing. The console hardware and procedure embody many features, worked out through the years, that maximize convenience and free the user from clerical routine. The formats and procedures are quite flexible. The user learns, through working with the system, what modes and techniques suit him best. Ordinarily, he gives the system rather terse, almost minimal instructions, relying on it to interpret them correctly and to do what he wishes and expects. When it misinterprets him or gets off the track of his thinking, as it sometimes does, he falls back on more explicit expression of commands and queries.

To continue with our example, let us move on to Monday. The reference citations and abstracts are ready for examination. The system has anticipated that I may want to see or process the full texts, and they are now available in secondary memory, having been moved up from the tertiary store along with a lot of other, somewhat less clearly relevant, material. I do not know exactly how much of such anticipatory preparation has gone on within the system, but I know that the pressure of on-line requests is low during the week-end, and I am counting on the system to have done a fair amount of work on my behalf. (I could have explicitly requested preparatory assembly and organization of relevant material, but it is much less expensive to let the system take the initiative. The system tends to give me a fairly high priority because I often contribute inputs intended to improve its capabilities.) Actually, the system has been somewhat behind schedule in its organization of information in the field of my interest, but over the week-end it retrieved over 10,000 documents, scanned them all for sections rich in relevant material, analyzed all the rich sections into statements in a high-order predicate calculus, and entered the statements into the data base of the question-answering subsystem. . . .

How [did] the system approach the problem of selecting relevant documents? The approach to be described is not advanced far beyond the actual state of the art in 1964. Certainly, a more sophisticated approach will be feasible before 1994.

All contributions to the system are assigned tentative descriptors when the contributions are generated. The system maintains an elaborate thesaurus of descriptors and related terms and expressions. The thesaurus recognizes many different kinds of relations between and among terms. It recognizes different meanings of a given term. It recognizes logical categories and syntactic categories. The system spends much time maintaining and improving this thesaurus. As soon as it gets a chance, it makes statistical analyses of the text of a new acquisition and checks the tentatively assigned descriptors against the analyses. It also makes a combined syntactic-semantic analysis of the text, and reduces every sentence to a (linguistic) canonical form and also to a set of expressions in a (logical) predicate calculus. If it has serious difficulty in doing any of these things, it communicates with the author or editor, asking help in solving its problem or requesting revision of the text. It tests each unexpectedly frequent term (word or unitary phrase) of the text against grammatical and logical criteria to determine its appropriateness for use as a descriptive term, draws up a set of working descriptors and subdescriptors, sets them into a descriptor structure, and, if necessary, updates the general thesaurus.*

In selecting documents that may be relevant to a retrieval prescription, the system first sets up a descriptor structure for the prescription. This structure includes all the terms of the prescription that are descriptors or subdescriptors at any level in the thesaurus. It includes, also, thesaurus descriptors and subdescriptors that are synonymous to, or related in any other definite way to, terms of the prescription that are not descriptors or subdescriptors in the thesaurus. All the logical relations and modulations of the prescription are represented in its descriptor structure.

The descriptor structure of a document is comparable to the descriptor structure of a prescription. The main task of the system in screening documents for relevance, therefore, is to measure the degrees of correlation or congruence that exist between various parts of the prescription's structure and corresponding parts, if they exist, of each document's structure. This is done by an algorithm (the object of intensive study during development of the system) that determines how much various parts of one structure have to be distorted to make them coincide with parts of another. The algorithm yields two basic measures for each significant coincidence: (1) degree of match, and (2) size of matching substructure. The system then goes back to determine, for each significant coincidence, (3) the amount of text associated with the matching descriptor structure. All three measures

---

* New entries into the general thesaurus are dated. They remain tentative until proven through use.

are available to the user. Ordinarily, however, he works with a single index, which he is free to define in terms of the three measures. When the user says "best references," the system selects on the basis of his index. If the user has not defined an index, of course, the system defines one for him, knowing his fields of interest, who his colleagues are, how they defined their indexes, and so forth. . . .

Our information technology is not yet capable of constructing a significant, practical system of the type we have been discussing. If it were generally agreed, as we think it should be, that such a system is worth striving for, then it would be desirable to have an implementation program. The first part of such a program should not concern itself directly with system development. It should foster advancement of relevant sectors of technology.*

Let us assume then—though without insisting—that it is in the interest of society to accelerate the advances. What particular things should be done?

One of the first things to do, according to our study, is to break down the barriers that separate the potentially contributory disciplines. Among the disciplines relevant to the development of procognitive systems are (1) the library sciences, including the part of information storage and retrieval associated with the field of documentation, (2) the computer sciences, including both hardware and software aspects and the part of information storage and retrieval associated with computing, (3) the system sciences, which deal with the whole spectrum of problems involved in the design and development of systems, and (4) the behavioral and social sciences, parts of which are somewhat (and should be more) concerned with how people obtain and use information and knowledge. (The foregoing is not, of course, an exhaustive list; it even omits mathematical linguistics and mathematical logic, both of which are fundamental to the analysis and transformation of recorded knowledge.)

A second fundamental step is to determine basic characteristics of the relevance network that interrelates the elements of the fund of knowledge. . . .

There is, therefore, a need for an effective, formal, analytical semantics. With such a tool, one might hope to construct a network in which every element of the fund of knowledge is connected to every other element to which it is significantly related. . . .

The most necessary hardware development appears to be in the area of memory. . . . Procognitive systems will pose requirements for very large memories and for advanced memory organizations. Unless an unexpected breakthrough reconciles fast random access with very large capacity, there

---

* Science is also involved, of course, but for the sake of brevity "technology" is used in a very broad sense in this part of the discussion.

will be a need for memories that effect various compromises between those desiderata. . . .

As soon as it is feasible, moreover, multiple-console computer systems should be brought into contact with libraries. Perhaps they should be connected first to the card catalogues. Then they should be used in the development of descriptor-based retrieval systems. Almost certainly, the most promising way to develop procognitive systems is to foster their evolution from multiple-console computer systems—to arrange things in such a way that much of the conceptual and software development will be carried out by substantive users of the systems. . . .

To what extent should the language employed in the organization, direction, and use of procognitive systems resemble natural languages such as English? That question requires much study. If the answer should be, "Very closely," the implementation will require much research. Indeed, much research on computer processing of natural language will be required in any event, for the text of the existing corpus is largely in the form of natural language, and the body of knowledge will almost surely have to be converted into some more compact form in the interests of economy of storage, convenience of organization, and effectiveness of retrieval. . . .

The trouble with English as a carrier of knowledge is the horrendous amount of calculating on a very large base of data that is expended just to decide which of several locally plausible interpretations of a simple statement is correct or was intended. If the greater part of man's capability is wasted in that kind of processing, he does not have enough left to achieve more significant goals. This conclusion is obvious when the processing of English text is attempted by a present-day computer. It is less obvious but probably just as true for people.

The higher-order language that we propose as an effective carrier for knowledge is a kind of unambiguous English. As long as changes of context are signaled explicitly within the language, no serious problem is introduced by dependence on context. (Indeed, dependence on context appears to be necessary for the achievement of efficiency in diverse special applications.) The proposed language would recognize most or all of the operations, modulations, and qualifications that are available in English. However, it would quantize the continuous variables and associate one term or structure unambiguously with each degree. Finally, the system in which the proposed language is to be implemented would enforce consistent use of names for substantives; it would monitor "collisions" among terms, ask authors for clarifications, and disallow new or conflicting uses of established symbols.

All this advocacy of unambiguous, high-order language may encounter the disdainful accusation, "You're just asking for ruly English!" However, the situation is more favorable now for a ruly version of English than it ever has been, and it will be fully ripe before the new language is likely to be developed. The situation will be ripe, not because people will be ready to adopt a new dialect, but because computers with large data bases will need the new dialect as an information-input language. The envisioned sequence is: from (1) the natural (technical) language of the journal article through a machine-aided editorial translation into (2) unambiguous English, and then through a purely machine transformation into (3) the language(s) of the computer or of the data base itself. At any rate, this is a plausible approach that deserves investigation.

The Computer's Public Image

ROBERT S. LEE

IBM Corporation

In thinking and speculating about man's future relationship to the computer, perhaps it is best to start with an examination of where we are today. In line with this, I'd like to present some research findings which show that the computer has *already* had an impact on man's personality in certain rather significant respects. Now, I realize that this may be difficult to believe, particularly when we consider that what most people know of computers is a kind of second-hand knowledge—primarily based on their exposure to newspaper stories and television events such as election broadcasts. By and large, few people have yet seen a computer first-hand and even fewer have had any direct controlling contact with it . In other words, for the vast majority of the population, the computer is only something they know *about*—they do not have any direct knowledge based on personal experience with the machines.

Yet there has been a psychological impact—a dim awareness that because of the computer, the world is somehow a different place and that man's position in this world is no longer the same.

Let's take a look at some research evidence based on 3,000 personal interviews with a scientifically selected cross-section of the American public.[1] This study was conducted in May 1963. Before starting the survey, we first did a considerable amount of intensive exploratory investigation to make sure that we would properly cover the great variety of attitudes,

Reprinted from Robert S. Lee, "The Computer's Public Image," *Datamation*, 12 (December 1966), pp. 33–34, 39, copyright by F. D. Thompson Publications, Inc. Used with permission of *Datamation*.

[1] The quantitative evidence for most of the findings discussed in this paper were presented in a technical presentation of this study delivered at the 19th annual conference of the American Assn. of Public Opinion Research in May 1964 at Excelsior Springs, Mo., and at the annual convention of the American Sociological Assn. held in Montreal in September of that year.

opinions and reactions that exist about the computer and its social implications. In what follows, I shall describe in simple qualitative terms some of the broad findings that emerged from this investigation.

One of the most striking things to be seen in the data is the extent to which computers have excited the popular imagination. Even though only a very small segment of the population has had any substantial direct contact with the machines or with the environment of data processing, there is an enormous amount of interest in the topic. To most people, the computer is still something remote from their everyday lives—it is something they only hear about and read about. Yet, they are quite aware of its existence and they are fascinated by it. They frequently describe the computer as "amazing," "fantastic," and "astounding," and many of them are intensely eager to find out more about these machines—how they work and what they can do. There is, of course, a widespread recognition that computers are extremely complicated—that you have to have a lot of education and specialized training to comprehend the details of how they work.

Although popular thinking about computers is often rather unsophisticated and oversimplified, we discovered that the basic problem-solving function of computers appears to be rather well understood. Computers are closely associated with arithmetic, a well-recognized general tool that has a multitude of uses. And, although it is thought of frequently in terms of payroll, accounting and other office and business applications, the computer is also associated with a variety of other functions such as scientific research, space guidance and military defense. There seems to be no one outstanding application which dominates public thinking about computers; it is properly understood to be a general-purpose problem-solving device.

There is widespread appreciation that the computer is beneficial—both now and even more so in the future. The computer is seen as an aid to business organizations, as a tool to help scientists speed up scientific progress, and as important to the man-in-space program.

This appreciation of the computer as a beneficial tool is highest among people who are familiar with the world of business, among those who have an interest in mechanical things, among people who have an optimistic outlook, and among those who are generally openminded and receptive to things that are new and different. And, as we might expect, it is the young college-educated people who are personally most interested in finding out more about computers and how they work.

POPULAR MISCONCEPTIONS

The computer is a symbol of change—of innovation—of the future. Those with the greatest stake in the future and those who are psychologically most open to it are the people with the greatest interest and enthusiasm about the machines.

While the prevailing attitude toward computers is that they are an exciting and beneficial tool of progress, there are some fairly common misunderstandings about it. And, I'm afraid that perhaps we in the computer industry, in our enthusiasm, may have even contributed to some of these misconceptions.

First, there is the notion that the computer can do anything—that it can solve any problem. Then there is the idea that the computer always gives the right answer. People tend to generalize from the computer's well-known computational accuracy to the idea that the computer is infallible in solving the larger problems for which it may be used. Another popular misconception is the anthropomorphic notion that the machine is some sort of autonomous entity—a kind of superbrain which thinks as humans do and which can provide instant solutions to highly complicated problems that the ordinary man cannot even begin to understand.

There is very little popular understanding of the role that people play in solving problems with computers. By and large, the public does not understand the importance of systems development, of model building, of the program, or of the programmer. They don't understand the concept "garbage in, garbage out." What they see is the machine itself—not the human effort that makes it possible for the computer to do its job.

All this, of course, is not only false; it is also dangerous, as it leads to an unhealthy sense of awe for the machine as all powerful, as incomprehensible, and as independent of man's will. This sort of thinking leads to a feeling of uneasiness about computers which, we have found, can exist side by side with the widespread sense of excitement about the great benefits of these machines.

What all this means is that there are two completely independent currents of thought and emotion about the computer in our culture. The computer as a beneficial tool of man is the mainstream viewpoint, the dominant perspective about computers. But there is, however, this secondary undercurrent of uneasiness which is primarily related to the notion that the computer is an autonomous thinking machine.

Close and detailed analysis of the data shows that the central focus of this uneasiness about computers is not the threat of automation and job displacement. The main source of anxiety is the idea that there is some sort of science-fiction machine which can perform the functions of human thinking—functions which were previously thought to be the unique province of the human mind. This is reacted to as a down-grading of humans, and engenders a feeling of inferiority in relation to the abilities of computers. This idea—this disquieting undercurrent of uneasiness which co-exists with the prevailing excitement and optimism about computers—is a counter-theme of strong emotional significance for people. The anthropomorphic notion of a machine which can possibly out-think man is not easy to assimilate or to live with. It suggests that man is less unique than he thought and that he is therefore somehow less important.

MAN'S IMAGE OF SELF

It has been said that the scientific revolution has resulted in a number of assaults on man's egocentric conception of himself. Copernicus showed that our world is not the center of the universe, Darwin showed that man is part of the same evolutionary stream as animals, and Freud showed that man is not fully the master of his own mind. We now find that the emergence of electronic computers is another challenge to man's self-concept.

But what of the future—what will happen to our conception of ourselves and our machines as computers become more widely used? With the future development of time-sharing and a continued reduction in the costs of computing, in the years ahead we can expect to witness a "user explosion": there will be a vast increase in the number of people with access to computer terminals and various peripheral devices. Some day in the future, most people who will have direct daily contact with computers will not be computer experts such as engineers and other data processing professionals. The "new users" will be a different breed—they will be executives, managers, secretaries, sales persons, students and so forth. For them, the computer will be an auxiliary device, an aid in their daily work. This is in sharp contrast to the situation today where most people in contact with the machines have jobs primarily oriented around computers and data processing itself.

We can anticipate that such an increase in public visibility and contact with the machines is bound to have an effect on popular attitudes. But what will this effect be? Dr. Ulrich Neisser has studied some 60 users of the Project MAC time-sharing facility at MIT and reports a marked absence of anthropomorphic notions or anxieties about the machine.[2] This leads him to believe that the uneasiness that exists about computers is simply a passing phase in our relationship to the machines—that as we come into closer everyday contact with them, we will come to accept them with psychological comfort as we do with so many other modern inventions and innovations.

My own data tend to bear out this idea that familiarity reduces the sense of uneasiness about computers. We find that when we look at non-technically trained college-educated people who are not in direct contact with the machines but who are indirectly exposed to the world of data processing and computers through their work or by knowing people who make use of the machines—such people are less likely to feel uneasy about computers than those of comparable background who have not had this type of exposure. In other words, in the more educated segment of the population, uneasiness about computers seems to be highest among people who are most remote from them; it decreases with greater familiarity with the world of data processing, and it seems to be virtually non-existent among people who have direct contact with the machines.

I am not very sure, however, that we can so easily extrapolate these findings to what will happen when we have the user explosion—especially when we consider the possible reactions of non-technical people to certain sophisticated applications that combine adaptive, heuristic programming with conversational mode interaction. I don't think that we will simply learn to "accept" the computer; rather, I think that we will eventually develop a new and broadened conception of man, himself, now that machines exist which he can use as an extension of his intellect.

2 Neisser, U., *"Computers as Tools and as Metaphors."* Address delivered at the Georgetown Univ. Conference on Cybernetics and Society, November 20, 1964.

In the long run, it will probably be the youngsters who grow up in a world of computers who will more truly understand the new relationship of man and the machine. They will not be handicapped by older concepts, they will not feel a sense of ego threat, and they will not be in some sort of identity crisis in relation to machines. Instead, I believe they will be eager to understand the new tools, to use them to extend their knowledge and to extend their freedom to build the kind of future they want. And, out of this, I think a new concept of man will develop—a concept of man which includes and which accepts his ability to amplify the powers of his mind.

BRAINS & INTELLECT

In the course of human evolution, millions of years ago, the hominoid brain went through a period of great enlargement in a relatively short span of time. Anthropologists used to believe that the development of tools took place after this period of rapid brain development—that an increase in man's ability to deal with his environment led to his becoming a tool-using animal. That latest evidence, I am told, contradicts this. It is now believed that the human brain went through its rapid period of enlargement some time *after* man started to use tools.

In commenting on this discovery, Jerome Bruner, president of the American Psychological Assn. at that time, pointed out that since this evolutionary development—that is, over the past 500-thousand-thousand years—the principal change in man has been alloplastic rather than autoplastic. "That is, he has changed by linking himself with new, external implementation systems rather than by any conspicuous change in morphology—that is 'evolution-by-prosthesis'. . . ."[3]

In other words, after the great biological leap which was itself made possible by his development of tools, man's ability to behave more intelligently and to be a more important significant force in his environment was advanced by the further development of tools as an extension of his muscles and of his senses.

Today we are developing tools of an even higher order—machines that will be an intrinsic part of our intellectual equipment as they directly magnify our intellectual capabilities for intelligent thought and action.[4] The major problem will be to learn how to use these new tools wisely for truly human purposes. And, in this effort, our focus should be on man—on man in his new identity as extended and enlarged by the machine, not on the machine itself as something separate from and apart from man.

[3] Bruner, J. S., "The Course of Cognitive Growth," *American Psychologist,* 1964, 19, 1–15.
[4] Chein, I., "On the Nature of Intelligence," *Journal of General Psychology,* 1945, 32, 111–126.

Social Philosophy and World-View

BRUCE MAZLISH

Massachusetts Institute of Technology

A famous cartoon in *The New Yorker* magazine shows a large computer with two scientists standing excitedly beside it. One of them holds in his hand the tape just produced by the machine, while the other gapes at the message printed on it. In clear letters, it says, "Cogito, ergo sum," the famous Cartesian phrase, "I think, therefore I am."

My next cartoon has not yet been drawn. It is a fantasy on my part. In it, a patient, wild of eye and hair on end, is lying on a couch in a psychiatrist's office talking to an analyst, who is obviously a machine. The analyst-machine is saying, "Of course I'm human—aren't you?"[1]

These two cartoons are a way of suggesting the threat which the in-

Reprinted from Bruce Mazlish, "The Fourth Discontinuity," *Technology and Culture,* 8 (January 1967), pp. 1–15, copyright by The University of Chicago Press. Used with permission of the author and The University of Chicago Press.

[1] After finishing the early drafts of this article, I secured unexpected confirmation of my "fantasy" concerning an analyst-machine (which is not, in itself, critical to my thesis). A story in the *New York Times,* March 12, 1965, reports that "a computerized typewriter has been credited with remarkable success at a hospital here in radically improving the condition of several children suffering an extremely severe form of childhood schizophrenia. . . . What has particularly amazed a number of psychiatrists is that the children's improvement occurred without psychotherapy; only the machine was involved. It is almost as much human as it is machine. It talks, it listens, it responds to being touched, it makes pictures or charts, it comments and explains, it gives information and can be set up to do all this in any order. In short, the machine attempts to combine in a sort of science-fiction instrument all the best of two worlds—human and machine. It is called an Edison Responsive Environment Learning System. It is an extremely sophisticated 'talking' typewriter (a cross between an analogue and digital computer) that can teach children how to read and write. . . . Dr. Campbell Goodwin speculates that the machine was able to bring the autistic children to respond because it eliminated humans as communication factors. Once the children were able to communicate, something seemed to unlock in their minds, apparently enabling them to carry out further normal mental activities that had eluded them earlier."

creasingly perceived continuity between man and the machine poses to us today. It is with this topic that I wish to deal now, approaching it in terms of what I shall call the "fourth discontinuity." In order, however, to explain what I mean by the "fourth discontinuity," I must first place the term in a historical context.

In the eighteenth lecture of his *General Introduction to Psychoanalysis,* originally delivered at the University of Vienna between 1915 and 1917, Freud suggested his own place among the great thinkers of the past who had outraged man's naïve self-love. First in the line was Copernicus, who taught that our earth "was not the center of the universe, but only a tiny speck in a world-system of a magnitude hardly conceivable." Second was Darwin, who "robbed man of his peculiar privilege of having been specially created, and relegated him to a descent from the animal world." Third, now, was Freud himself. On his own account, Freud admitted, or claimed, that psychoanalysis was "endeavoring to prove to the 'ego' of each one of us that he is not even master in his own house, but that he must remain content with the veriest scraps of information about what is going on unconsciously in his own mind."

A little later in 1917, Freud repeated his sketch concerning the three great shocks to man's ego. In his short eassy, "A Difficulty in the Path of Psychoanalysis," he again discussed the cosmological, biological, and now psychological blows to human pride and, when challenged by his friend Karl Abraham, admitted, "You are right in saying that the enumeration of my last paper may give the impression of claiming a place beside Copernicus and Darwin."[2]

There is some reason to believe that Freud may have derived his conviction from Ernst Haeckel, the German exponent of Darwinism, who in his book *Natürliche Schöpfungsgeschichte* (1889) compared Darwin's achievement with that of Copernicus and concluded that together they had helped remove the last traces of anthropomorphism from science.[3] Whatever the origin of Freud's vision of himself as the last in the line of ego-shatterers, his assertion has been generally accepted by those, like Ernest Jones, who refer to him as the "Darwin of the Mind."[4]

The most interesting extension of Freud's self-view, however, has come from the American psychologist, Jerome Bruner. Bruner's version of what Freud called his "transvaluation" is in terms of the elimination of discontinuities, where discontinuity means an emphasis on breaks or gaps in the phenomena of nature—for example, a stress on the sharp differences between physical bodies in the heavens or on earth or between one form of animal matter and another—instead of an emphasis on its continuity. Put the other way, the elimination of discontinuity, that is, the establishment

[2] Ernest Jones, *The Life and Work of Sigmund Freud* (3 vols.; New York, 1953–57), II, 224–26.
[3] Ernst Cassirer, *The Problem of Knowledge: Philosophy, Science, and History since Hegel,* trans. William H. Woglom and Charles W. Hendel (New Haven, Conn., 1950), p. 160.
[4] Jones, III, 304.

of a belief in a continuum of nautre, can be seen as the creation of continuities, and this is the way Bruner phrases it. According to Bruner, the first continuity was established by the Greek physicist-philosophers of the sixth century, rather than by Copernicus. Thus, thinkers like Anaximander conceived of the phenomena of the physical world as "continuous and monistic, as governed by the common laws of matter."[5] The creating of the second continuity, that between man and the animal kingdom was, of course, Darwin's contribution, a necessary condition for Freud's work. With Freud, according to Bruner, the following continuities were established: the continuity of organic lawfulness, so that "accident in human affairs was no more to be brooked as 'explanation' than accident in nature"; the continuity of the primitive, infantile, and archaic as co-existing with the civilized and evolved; and the continuity between mental illness and mental health.

In this version of the three historic ego-smashings, man is placed on a continuous spectrum in relation to the universe, to the rest of the animal kingdom, and to himself. He is no longer discontinuous with the world around him. In an important sense, it can be contended, once man is able to accept this situation, he is in harmony with the rest of existence. Indeed, the longing of the early nineteenth-century romantics, and of all "alienated" beings since, for a sense of "connection" is fulfilled in an unexpected manner.

Yet, to use Bruner's phraseology, though not his idea, a fourth and major discontinuity, or dichotomy, still exists in our time. It is the discontinuity between man and machine. In fact, my thesis is that this fourth discontinuity must now be eliminated—indeed, we have started on the task—and that in the process man's ego will have to undergo another rude shock, similar to those administered by Copernicus (or Galileo), Darwin, and Freud. To put it bluntly, we are now coming to realize that man and the machines he creates are continuous and that the same conceptual schemes, for example, that help explain the workings of his brain also explain the workings of a "thinking machine." Man's pride, and his refusal to acknowledge this continuity, is the substratum upon which the distrust of technology and an industrialized society has been reared. Ultimately, I believe, this last rests on man's refusal to understand and accept his own nature—as a being continuous with the tools and machines he constructs. Let me now try to explain what is involved in this fourth discontinuity.

*　　*　　*

5 For Bruner's views, see his "Freud and the Image of Man," *Partisan Review,* **XXIII**, No. 3 (Summer 1956), 340–47. In place of both Bruner's sixth-century Greek physicists and Freud's Copernicus, I would place Galileo as the breaker of the discontinuity that was thought to exist in the material world. It was Galileo, after all, who first demonstrated that the heavenly bodies are of the same substance as the "imperfect" earth and subject to the same mechanical laws. In his *Dialogue on the Two Principal World Systems* (1632), he not only supported the "world system" of Copernicus against Ptolemy but established that our "world," i.e., the earth, is a natural part of the other "world," i.e., the solar system. Hence, the universe at large is one "continuous" system, a view at best only implied in Copernicus. Whatever the correct attribution, Greek physicists, Copernicus, or Galileo, Freud's point is not in principle affected.

The evidence seems strong today that man evolved from the other animals into humanity through a continuous interaction of tool, physical, and mental-emotional changes. The old view that early man arrived on the evolutionary scene, fully formed, and then proceeded to discover tools and the new ways of life which they made possible is no longer acceptable. As Sherwood L. Washburn, professor of anthropology at the University of California, puts it, "From the rapidly accumulating evidence it is now possible to speculate with some confidence on the manner in which the way of life made possible by tools changed the pressures of natural selection and so changed the structure of man." The details of Washburn's argument are fascinating, with its linking of tools with such physical traits as pelvic structure, bipedalism, brain structure, and so on, as well as with the organization of men in co-operative societies and the substitution of morality for hormonal control of sexual and other "social" activities. Washburn's conclusion is that "it was the success of the simplest tools that started the whole trend of human evolution and led to the civilizations of today."[6]

Darwin, of course, had had a glimpse of the role of tools in man's evolution.[7] It was Karl Marx, however, who first placed the subject in a new light. Accepting Benjamin Franklin's definition of man as a "tool-making animal," Marx suggested in *Das Kapital* that "the relics of the instruments of labor are of no less importance in the study of vanished socio-economic forms, than fossil bones are in the study of the organization of extinct species." As we know, Marx wished to dedicate his great work to Darwin—a dedication rejected by the cautious biologist—and we can see part of Marx's reason for this desire in the following revealing passage:

> Darwin has aroused our interest in the history of *natural technology,* that is to say in the origin of the organs of plants and animals as productive instruments utilised for the life purposes of those creatures. Does not the history of the origin of the productive organs of men in society, the organs which form the material basis of every kind of social organisation, deserve equal attention? Since, as Vico [in the *New Science* (1725)] says, the essence of the distinction between human history and natural history is that the former is the work of man and the latter is not, would not the history of *human technology* be easier to write than the history of natural technology? Technology reveals man's dealings with nature, discloses the direct productive activities of his life, thus throwing light upon social relations and the resultant mental conceptions.[8]

Only a dogmatic anti-Marxist could deny that Marx's brilliant imagination had led him to perceive a part of the continuity between man and his tools. Drawn off the track, perhaps, by Vico's distinction between human and natural history as man-made and God-made, Marx might almost be given a place in the pantheon of Copernicus, Darwin, and Freud as a

---

6 "Tools and Human Evolution," *Scientific American,* CCIII, No. 3 (September 1960), 63–75.
7 E.g., see Charles Darwin, *The Descent of Man* (New York, n.d.), pp. 431–32, 458.
8 Italics mine; Karl Marx, *Capital,* trans. Eden and Cedar Paul (2 vols.; London, 1951), I, 392–93, n. 2.

destroyer of man's discontinuities with the world about him. Before our present-day anthropologists, Marx had sensed the unbreakable connection between man's evolution as a social being and his development of tools. He did not sense, however, the second part of our subject, that man and his tools, especially in the form of modern, complicated machines, are part of a theoretical continuum.

*   *   *

The *locus classicus* of the modern insistence on the fourth discontinuity is, as is well known, the work of Descartes. In his *Discourse on Method,* for example, he sets up God and the soul on one side, as without spatial location or extension, and the material-mechanical world in all its aspects, on the other side. Insofar as man's mind or soul participates in reason—which means God's reason—man knows this division or dualism of mind and matter, for, as Descartes points out, man could not know this fact from his mere understanding, which is based solely on his senses, "a location where it is clearly evident that the ideas of God and the soul have never been."[9]

Once having established his God, and man's participation through reason in God, Descartes could advance daringly to the very precipice of a world without God. He conjures up a world in imaginary space and shows that it must run according to known natural laws. Similarly, he imagines that "God formed the body of a man just like our own, both in the external configuration of its members and in the internal configuration of its organs, without using in its composition any matter but that which I had described [i.e., physical matter]. I also assumed that God did not put into this body any rational soul [defined by Descartes as "that part of us distinct from the body whose essence...is only to think"]."

Analyzing this purely mechanical man, Descartes boasts of how he has shown "what changes must take place in the brain to cause wakefulness, sleep, and dreams; how light, sounds, odors, taste, heat, and all the other qualities of external objects can implant various ideas through the medium of the senses...I explained what must be understood by that animal sense which receives these ideas, by memory which retains them, and by imagination which can change them in various ways and build new ones from them." In what way, then, does such a figure differ from real man? Descartes confronts his own created "man" forthrightly; it is worth quoting the whole of his statement:

> Here I paused to show that if there were any machines which had the organs and appearance of a monkey or of some other unreasoning animal, we would have no way of telling that it was not of the same nature as these animals. But if there were a machine which had such a resemblance to our bodies, and imitated our actions as far as possible, there would always be two absolutely certain methods of recognizing that it was still not truly a man. The

[9] René Descartes, *Discourse on Method,* trans. Laurence J. Lafleur (Indianapolis, 1956), p. 24. The rest of the quotations are also from this translation, pp. 29, 35–36, and 36–37.

first is that it could never use words or other signs for the purpose of communicating its thoughts to others, as we do. It indeed is conceivable that a machine could be so made that it would utter words, and even words appropriate to physical acts which cause some change in its organs; as, for example, if it was touched in some spot that it would ask what you wanted to say to it; if in another, that it would cry that it was hurt, and so on for similar things. But it could never modify its phrases to reply to the sense of whatever was said in its presence, as even the most stupid men can do. The second method of recognition is that although such machines could do many things as well as, or perhaps even better than, men, they would infallibly fail in certain others, by which we would discover that they did not act by understanding, but only by the disposition of their organs. For while reason is a universal instrument which can be used in all sorts of situations, the organs have to be arranged in a particular way for each particular action. From this it follows that it is morally impossible that there should be enough different devices in a machine to make it behave in all the occurrences of life as our reason makes us behave.

Put in its simplest terms, Descartes' two criteria for discriminating between man and the machine are that the latter has (1) no feedback mechanism ("it could never modify its phrases") and (2) no generalizing reason ("reason is a universal instrument which can be used in all sorts of situations"). But it is exactly in these points that, today, we are no longer able so surely to sustain the dichotomy. The work of Norbert Wiener and his followers, in cybernetics, indicates what can be done on the problem of feedback. Investigations into the way the brain itself forms concepts are basic to the attempt to build computers that can do the same, and the two efforts are going forward simultaneously, as in the work of Dr. W. K. Taylor of University College, London, and of others. As G. Rattray Taylor sums up the matter: "One can't therefore be quite as confident that computers will one day equal or surpass man in concept-forming ability as one can about memory, since the trick hasn't yet been done; but the possibilities point that way."[10] In short, the gap between man's thinking and that of his thinking machines has been greatly narrowed by recent research.

Descartes, of course, would not have been happy to see such a development realized. To eliminate the dichotomy or discontinuity between man and machines would be, in effect, to banish God from the universe. The rational soul, Descartes insisted, "could not possibly be derived from the powers of matter...but must have been specially created." Special creation requires God, for Descartes' reasoning is circular. The shock to man's ego, of learning the Darwinian lesson that he was not "specially created," is, in this light, only an outlying tremor of the great earthquake that threatened man's view of God as well as of himself. The obstacles to removing not only the first three but also the fourth discontinuity are, clearly, deeply imbedded in man's pride of place.

[10] See G. Rattray Taylor, "The Age of the Androids," *Encounter* (November 1963), p. 43. On p. 40 Taylor gives some of the details of the work of W. K. Taylor and others.

How threatening these developments were can be seen in the case of Descartes' younger contemporary, Blaise Pascal. Aware that man is "a thinking reed," Pascal also realized that he was "engulfed in the infinite immensity of spaces whereof I know nothing, and which knows nothing of me." "I am terrified," he confessed. To escape his feeling of terror, Pascal fled from reason to faith, convinced that reason could not bring him to God. Was he haunted by his own construction, at age nineteen, of a calculating machine which, in principle, anticipated the modern digital computer? By his own remark that "the arithmetical machine produces effects which approach nearer to thought than all the actions of animals"? Ultimately, to escape the anxiety that filled his soul, Pascal commanded, "On thy knees, powerless reason."[11]

Others, of course, walked where angels feared to tread. Thus, sensationalist psychologists and epistemologists, like Locke, Hume, or Condillac, without confronting the problem head on, treated the contents of man's reason as being formed by his sense impressions. Daring thinkers, like La Mettrie in his *L'Homme machine* (1747) and Holbach, went all the way to a pure materialism. As La Mettrie put it in an anticipatory transcendence of the fourth discontinuity, "I believe thought to be so little incompatible with organized matter that it seems to be a property of it, like Electricity, Motive Force, Impenetrability, Extension, etc."[12]

On the practical front, largely leaving aside the metaphysical aspects of the problem, Pascal's work on calculating machines was taken up by those like the eccentric nineteenth-century mathematician Charles Babbage, whose brilliant designs outran the technology available to him.[13] Thus it remained for another century, the twentieth, to bring the matter to a head and to provide the combination of mathematics, experimental physics, and modern technology that created the machines that now confront us and that reawaken the metaphysical question.

*   *   *

The implications of the metaphysical question are clear. Man feels threatened by the machine, that is, by his tools writ large, and feels out of harmony with himself because he is out of harmony—what I have called discontinuous—with the machines that are part of himself. Today, it is fashionable to describe such a state by the term "alienation." In the Marxist phraseology, we are alienated from ourselves when we place false gods or economics over us and then behave as if they had a life of their own, external and independent of ourselves, and, indeed, in control of our lives.

11 For details, see J. Bronowski and Bruce Mazlish, *The Western Intellectual Tradition: From Leonardo to Hegel* (New York, 1960), pp. 233–41.
12 See Stephen Toulmin, "The Importance of Norbert Wiener," *New York Review of Books,* September 24, 1964, p. 4, for an indication of La Mettrie's importance in this development. While Toulmin does not put his material in the context of the fourth discontinuity, I find we are in fundamental agreement about what is afoot in this matter.
13 See Philip and Emily Morrison (eds.), *Charles Babbage and His Calculating Engines* (New York, 1961).

My point, while contact can be established between it and the notion of alienation, is a different one. It is in the tradition of Darwin and Freud, rather than of Marx, and is concerned more with man's ego than with his sense of alienation.

A brief glimpse at two "myths" concerning the machine may illuminate what I have in mind. The first is Samuel Butler's negative utopia, *Erewhon,* and the second is Mary Shelley's story of Frankenstein. In Butler's novel, published in 1872, we are presented with Luddism carried to its final point. The story of the Erewhonian revolution against the machines is told in terms of a purported translation from a manuscript, "The Book of the Machines," urging men on to the revolt and supposedly written just before the long civil war between the machinists and the antimachinists, in which half the population was destroyed. The prescient flavor of the revolutionary author's fears can be caught in such passages as follows: [14]

> "There is no security"—to quote his own words—"against the ultimate development of mechanical consciousness, in the fact of machines possessing little consciousness now. A mollusc has not much consciousness. Reflect upon the extraordinary advance which machines have made during the last few hundred years, and note how slowly the animal and vegetable kingdoms are advancing. The more highly organized machines are creatures not so much of yesterday as of the last five minutes, so to speak, in comparison with past time. Assume for the sake of argument that conscious beings have existed for some twenty million years: see what strides machines have made in the last thousand! May not the world last twenty million years longer? If so, what will they not in the end become? Is it not safer to nip the mischief in the bud and to forbid them further progress?
>
> "But who can say that the vapour-engine has not a kind of consciousness? Where does consciousness begin, and where end? Who can draw the line? Who can draw any line? Is not everything interwoven with everything? Is not machinery linked with animal life in an infinite variety of ways? The shell of a hen's egg is made of a delicate white ware and is a machine as much as an egg-cup is; the shell is a device for holding the egg as much as the egg-cup for holding the shell: both are phases of the same function; the hen makes the shell in her inside, but it is pure pottery. She makes her nest outside of herself for convenience' sake, but the nest is not more of a machine than the egg-shell is. A 'machine' is only a 'device.' "

Then he continues:

> "Do not let me be misunderstood as living in fear of any actually existing machine; there is probably no known machine which is more than a prototype of future mechanical life. The present machines are to the future as the early Saurians to man. The largest of them will probably greatly diminish in size. Some of the lowest vertebrata attained a much greater bulk than has descended to their more highly organized living representatives, and in like manner a diminution in the size of machines has often attended their development and and progress."

[14] The quotations that follow are from Samuel, Butler, *Erewhon* (Baltimore, 1954), pp. 161, 164, 167–68, and 171.

Answering the argument that the machine, even when more fully developed, is merely man's servant, the writer contends:

"But the servant glides by imperceptible approaches into the master; and we have come to such a pass that, even now, man must suffer terribly on ceasing to benefit the machines.... Man's very soul is due to the machines; it is a machine-made thing; he thinks as he thinks, and feels as he feels, through the work that machines have wrought upon him, and their existence is quite as much a *sine qua non* for his, as his for theirs. This fact precludes us from proposing the complete annihilation of machinery, but surely it indicates that we should destroy as many of them as we can possibly dispense with, lest they should tyrannize over us even more completely."

And, finally, the latent sexual threat is dealt with:

"It is said by some with whom I have conversed upon this subject, that the machines can never be developed into animate or quasi-animate existences, inasmuch as they have no reproductive systems, nor seem ever likely to possess one. If this be taken to mean that they cannot marry, and that we are never likely to see a fertile union between two vapour-engines with the young ones playing about the door of the shed, however greatly we might desire to do so, I will readily grant it. But the objection is not a very profound one. No one expects that all the features of the now existing organizations will be absolutely repeated in an entirely new class of life. The reproductive system of animals differs widely from that of plants, but both are reproductive systems. Has nature exhausted her phases of this power?"

Inspired by fears such as these, which sound like our present realities, the Erewhonians rise up and destroy almost all their machines. It is only years after this supposed event that they are sufficiently at ease so as to collect the fragmentary remains, the "fossils," of the now defunct machines and place them in a museum. At this point, the reader is never sure whether Butler's satire is against Darwin or the anti-Darwinists, probably both, but there is no question of the satire when he tells us how machines were divided into "their genera, subgenera, species, varieties, subvarieties, and so forth" and how the Erewhonians "proved the existence of connecting links between machines that seemed to have very little in common, and showed that many more such links had existed, but had now perished." It is as if Butler had taken Marx's point about *human technology* and stood it on its head!

Going even further, Butler foresaw the threatened ending of the fourth discontinuity, just as he saw Darwin's work menacing the third of the discontinuities we have discussed. Thus, we find Butler declaring, in the guise of his Erewhonian author, "I shrink with as much horror from believing that my race ever be superseded or surpassed, as I should do from believing that even at the remotest period my ancestors were other than human beings. Could I believe that ten hundred thousand years ago a single one of my ancestors was another kind of being to myself, I should lose all self-respect, and take no further pleasure or interest in life. I have the same feeling with regard to my descendants and believe it to be one that will be felt so generally that the country will resolve upon putting an immediate stop to all further mechanical progress, and upon destroying all improvements that have been made for the last three hundred years." The counter-argument, that "machines were to be regarded as a part of man's own physical nature, being really nothing but extra-corporeal limbs. Man [is] a machinate mammal," is dismissed out of hand.

Many of these same themes—the servant-machine rising against its master, the fear of the machine reproducing itself (fundamentally, a sexual fear, as Caliban illustrates, and as our next example will show), the terror, finally, of man realizing that he is at one with the machine—can be found attached to an earlier myth, that of Frankenstein. Now passed into our folklore, people frequently give little attention to the actual details of the novel. First, the name Frankenstein is often given to the monster created, rather than to its creator; yet, in the book, Frankenstein is the name of the scientist, and his abortion *has no name*. Second, the monster is *not* a machine but a "flesh and blood" product; even so informed a student as Oscar Handlin makes the typical quick shift, in an echo of Butler's fears, when he says, "The monster, however, quickly proves himself the superior. In the confrontation, the machine gives the orders."[15] Third, and last, it is usually forgotten or overlooked that the monster turns to murder *because* his creator, horrified at his production, refuses him human love and kindness. Let us look at a few of the details.

In writing her "Gothic" novel in 1816–17, Mary Shelley gave it the subtitle, "The Modern Prometheus."[16] We can see why if we remember that Prometheus defied the gods and gave fire to man. Writing in the typical early nineteenth-century romantic vein, Mary Shelley offers Frankenstein as an example of "how dangerous is the acquirement of knowledge"; in this case, specifically, the capability of "bestowing animation upon lifeless matter." In the novel we are told of how, having collected his materials from "the dissecting room and the slaughter-house" (as Wordsworth has said of modern science, "We murder to dissect"), Frankenstein eventually completes his loathsome task when he infuses "a spark of being into the lifeless thing that lay at my feet." Then, as he tells us, "now that I had finished, the beauty of the dream vanished, and breathless horror and disgust filled my heart." Rushing from the room, Frankenstein goes to his bedchamber, where he has a most odd dream concerning the corpse of his dead mother—the whole book as well as this passage cries out for psychoanalytic interpretation—from which he is awakened by "the wretch—the miserable monster whom I had created." Aghast at the countenance of what he has created, Frankenstein escapes from the room and out into the open. Upon finally returning to his room with a friend, he is relieved to find the monster gone.

To understand the myth, we need to recite a few further details in this weird, and rather badly written, story. Frankenstein's monster eventually

15 "Science and Technology in Popular Culture," *Daedalus* (Winter 1965), p. 163.
16 The quotations that follow are from Mary Shelley, *Frankenstein* (New York, 1953), pp. 30–31, 36–37, 85, and 160–61.

finds his way to a hovel attached to a cottage, occupied by a blind father and his son and daughter. Unperceived by them, he learns the elements of social life (the fortuitous ways in which this is made to occur may strain the demanding reader's credulity), even to the point of reading *Paradise Lost*. Resolved to end his unbearable solitude, the monster, convinced that his virtues of the heart will win over the cottagers, makes his presence known. The result is predictable: horrified by his appearance, they duplicate the behavior of his creator, and flee. In wrath, the monster turns against the heartless world. He kills, and his first victim, by accident, is Frankenstein's young brother.

Pursued by Frankenstein, a confrontation between creator and created takes place, and the monster explains his road to murder. He appeals to Frankenstein in a torrential address:

> "I entreat you to hear me, before you give vent to your hatred on my devoted head. Have I not suffered enough that you seek to increase my misery? Life, although it may only be an accumulation of anguish, is dear to me, and I will defend it. Remember, thou hast made me more powerful than thyself; my height is superior to thine; my joints more supple. But I will not be tempted to set myself in opposition to thee. I am thy creature, and I will be even mild and docile to my natural lord and king, if thou wilt also perform thy part, the which thou owest me. Oh, Frankenstein, be not equitable to every other, and trample upon me alone, to whom thy justice, and even thy clemency and affection is most due. Remember, that I am thy creature; I ought to be thy Adam; but I am rather the fallen angel, whom thou drives from joy for no misdeed. Everywhere I see bliss, from which I alone am irrevocably excluded. I was benevolent and good; misery made me a fiend. Make me happy, and I shall again be virtuous."

Eventually, the monster extracts from Frankenstein a promise to create a partner for him "of another sex," with whom he will then retire into the vast wilds of South America, away from the world of men. But Frankenstein's "compassion" does not last long. In his laboratory again, Frankenstein indulges in a long soliloquy:

> "I was now about to form another being, of whose dispositions I was alike ignorant; she might become ten thousand times more malignant than her mate; and delight, for its own sake, in murder and wretchedness. He had sworn to quit the neighborhood of man, and hide himself in deserts; but she had not; and she, who in all probability was to become a thinking and reasoning animal, might refuse to comply with a compact made before her creation. They might even hate each other; the creature who already lived loathed his own deformity, and might he not conceive a greater abhorrence for it when it came before his eyes in the female form? She also might turn with disgust from him to the superior beauty of man; she might quit him, and he be again alone, exasperated by the fresh provocation of being deserted by one of his own species.
>
> "Even if they were to leave Europe, and inhabit the deserts of the new world, yet one of the first results of those sympathies for which the demon thirsted would be children, and a race of devils would be propagated upon

the earth who might make the very existence of the species of man a condition precarious and full of terror. Had I right, for my own benefit, to inflict this curse upon everlasting generations?"

With the monster observing him through the window, Frankenstein destroys the female companion on whom he had been working. With this, the novel relentlessly winds its way to its end. In despair and out of revenge, the monster kills Frankenstein's best friend, Clerval, then Frankenstein's new bride, Elizabeth. Fleeing to the frozen north, the monster is tracked down by Frankenstein (shades of Moby Dick?), who dies, however, before he can destroy him. But it does not matter; the monster wishes his own death and promises to place himself on a funeral pile and thus at last secure the spiritual peace for which he has yearned.

I have summarized the book because I suspect that few readers will actually be acquainted with the myth of Frankenstein *as written* by Mary Shelley. For most of us, Frankenstein is Boris Karloff, clumping around stiff, automatic, and threatening: a machine of sorts. We shall have forgotten completely, if ever we knew, that the monster, *cum* machine, is evil, or rather, becomes evil, only because it is spurned by man.

*   *   *

My thesis has been that man is on the threshold of breaking past the discontinuity between himself and machines. In one part, this is because man now can perceive his own evolution as inextricably interwoven with his use and development of tools, of which the modern machine is only the furthest extrapolation. We cannot think any longer of man without a machine. In another part, this is because modern man perceives that the same scientific concepts help explain the workings of himself and of his machines and that the evolution of matter—from the basic building blocks of hydrogen turning into helium in the distant stars, then fusing into carbon nuclei and on up to iron, and then exploding into space, which has resulted in our solar system—continues on earth in terms of the same carbon atoms and their intricate patterns into the structure of organic life. And now into the architecture of our thinking machines.

It would be absurd, of course, to contend that there are no differences between man and machines. This would be the same *reductio ad absurdum* as involved in claiming that, because he is an animal, there is no difference between man and the other animals. The matter, of course, is one of degree.[17] What is claimed here is that the sharp discontinuity between man and machines is no longer tenable, in spite of the shock to our egos.

---

[17] In semi-facetious fashion, I have argued with some of my more literal-minded friends that what distinguishes man from existing machines, and probably will always so distinguish him, is an *effective* Oedipus complex: *vive la différence!* For an excellent and informed philosophical treatment of the difference between man and machines, see J. Bronowski, *The Identity of Man* (Garden City, N.Y., 1965).

Scientists, today, know this; the public at large does not, *New Yorker* cartoons to the contrary.[18]

Moreover, this change in our metaphysical awareness, this transcendence of the fourth discontinuity, is essential to our harmonious acceptance of an industrialized world. The alternatives are either a frightened rejection of the "Frankensteins" we have created or a blind belief in their "superhuman virtues" and a touching faith that they can solve all our human problems. Alas, in the perspective I have suggested, machines are mechanical, all too mechanical," to paraphrase Nietzsche. But, in saying this, I have already also said that they are "all too human" as well. The question, then, is whether we are to repeat the real Frankenstein story and, turning from the "monsters" we have created, turn aside at the same time from our own humanity or, alternatively, whether we are to accept the blow to our egos and enter into a world beyond the fourth discontinuity?

[18] As in so much else, children "know" what their parents have forgotten. As O. Mannoni tells us, in the course of explaining totemism, "children, instead of treating animals as machines, treat machines as living things, the more highly prized because they are easier to appropriate. Children's appropriation is a virtual identification and they play at being machines (steam-engines, motor cars, aeroplanes) just as 'primitive' people play at being the totem [animal]" (*Prospero and Caliban, The Psychology of Colonization,* trans. Pamela Powesland [New York, 1964], p. 82). In *Huckleberry Finn,* Mark Twain puts this "identification" to work in describing Tom Sawyer's friend, Ben Rogers: "He was eating an apple, and giving a long, melodious whoop, at intervals, followed by a deep-toned ding-dong-dong, ding-dong-dong, for he was personating a steamboat. As he drew near, he slackened speed, took the middle of the street, leaned far over to starboard and rounded to ponderously and with laborious pomp and circumstance—for he was personating the *Big Missouri,* and considered himself to be drawing nine feet of water. He was boat and captain and engine-bells combined, so he had to imagine himself standing on his own hurricane-deck giving the orders and executing them!...'Stop the stabboard! Ting-a-ling-ling! Stop the labboard! Come ahead on the stabboard! Stop her! Let your outside turn over slow! Ting-a-ling! Chow-ow-aw! Get out that head-line! *Lively* now! Come—out with your spring-line—what're you about there! Take a turn round that stump with the bight of it! Stand by that stage, now—let her go! Done with the engines, sir! Ting-a-ling! sh't! sh't! sh't!' " See the analysis of this passage in Erik H. Erikson, *Childhood and Society* (2d ed.; New York, 1963), pp. 209 ff.

The Computer as Seen Through Science Fiction

MARCIA ASCHER

Ithaca College

Computers have become an intrinsic part of our environment. In magazines and books ample coverage is given to their efficiency, their broad applicability, and their profitability. An equally important topic, but one less frequently discussed, is the responsibilities accompanying the "computer revolution." A stimulating source of such discussions is science fiction stories. In their traditional role of insightful social and scientific gadflies, these stories provide enjoyment as well as focus attention on the relationship of men and computers. The stories influence popular opinion, too. Executives interested in the changing moods of the public would do well to keep an eye on what the science fiction writers are saying. The fears and problems dramatized in their spell-binders go into the minds of many millions of people.

My purpose in this article is not to evaluate or criticize science fiction concerning the computer. It is rather to sample the literature and briefly describe some of the leading themes that might interest businessmen. I shall use a broad canvas, including a couple of nineteenth century classics, contemporary computer stories from a variety of business and nonbusiness settings, and even a sample of Soviet science fiction.

*　　*　　*

One of the earliest themes in science fiction stories dealing with mechanization was that machines might develop consciousness and turn on their makers. As early as 1872 this idea was expressed in *Erewhon* by Samuel

Reprinted from Marcia Ascher, "Computers in Science Fiction," *Harvard Business Review,* 41 (November-December 1963), pp. 40–45, © by The President and Fellows of Harvard College. Used with permission of the author and the *Harvard Business Review.*

Butler.[1] A philosopher of Erewhon argues that by evolution machines may become conscious, enslave man, and supersede him. He contends that machines already have many advantages over men:

"...take man's vaunted power of calculation. Have we not engines which can do all manner of sums more quickly and correctly than we can?...In fact, wherever precision is required man flies to the machine at once, as far preferable to himself. Our sum-engines never drop a figure...the machine is brisk and active, when the man is weary; it is clear-headed and collected, when the man is stupid and dull; it needs no slumber, when man must sleep or drop; ever at its post, ever ready for work, its alacrity never flags, its patience never gives in."

The philosopher of Erewhon foresees that man will someday stand in the same relation to machines as domesticated animals now stand to man, but he adds:

"There is reason to hope that the machines will use us kindly, for their existence will be in a great measure dependent upon ours; they will rule us with a rod of iron, but they will not eat us; they will not only require our services in the reproduction and education of their young, but also in waiting upon them as servants; in gathering food for them, and feeding them; in restoring them to health when they are sick; and in either buying their dead or working up their deceased members into new forms of mechanical existence."

As a result of his argument, all the machines in Erewhon are completely destroyed.

Man fares less well when the tables are turned in another early story, "Moxon's Master," written by Ambrose Bierce in 1893.[2] In this tale an automated chess player, angry at being checkmated, strangles its maker. With less intensity, the same theme appears in a modern Soviet science fiction story, "Spontaneous Reflex."[3] A computer called Urm is built to be used in interplanetary travel to investigate and collect data from unknown localities. To ensure that the computer can function in any terrain or eventuality, a program that generates programs is written to guide the machine. We are informed that:

"Within the framework of the possibilities allowed by the basic programme, [the computer] would investigate this new thing, and either overcome it, if it could be overcome, or use it in the interests of the basic programme. That is to say, Urm would select without the help of men the most suitable type of behavior for any new event."

Somehow, one night in the laboratory, the self-programing capability creates such a need within the computer to investigate its environment that it initiates its own motion. Before it has satisfied its curiosity, the computer arouses fear, causes damage, and has to be destroyed.

[1] New York, New American Library of World Literature, Inc., 1961.
[2] *The Collected Writings of Ambrose Bierce* (New York, Citadel Press, 1946), p. 429.
[3] Arkady and Boris Strugatsky, *Soviet Science Fiction* (New York, Crowell-Collier Publishing Company, 1962), p. 89.

The pessimistic notion that man may create a machine which can get beyond his control and is capable of destroying him is closely allied to the more general science fiction theme that through the misuse of technology mankind will destroy itself. This theme certainly reflects an existent public fear, and its appearance centers about other technological devices as well as computers. While these pessimistic views do exist in science fiction stories, they are indeed in the minority. Most of the authors would seem more inclined to agree with an editorial in a local newspaper consoling "those who may have uneasy feelings that machines may one day take over the world" with the reminder that "one can always pull out the plug."[4]

Rather than emphasize physical dangers or catastrophe, science fiction plays up the moral responsibilities that accompany technological advancement. In particular, the stories involving computers stress that the ultimate responsibility for creativeness and decisions belongs to man.

*   *   *

Frequently computers appear in science fiction stories where they are not essential to the plots. They are then only another aspect of the distant time or place being described. In many of these cases the use and description of the computer are either farfetched or vague. But in most of the serious narratives this is not true. In fact, the application of the computer is clearly an extension of a way in which it is being applied today or an application that is currently being considered. What is happening today is used as a basis for visualizing what will take place tomorrow. By placing the action of the stories in the future, the problems raised by computer usage are isolated and examined, and the implications of present trends are highlighted.

One common theme is that man must exercise great care in giving instructions to a computer because of the literal way in which the instructions are interpreted. An example of a story with this as its central theme is "They've Been Working On. . . ."[5] Here a computer is used for the routing and scheduling of a railroad, an application that is already initiated. The intent of the computer system is "to keep track of all orders and loads and everything that travels by rail. . . and make every shipment reach its destination on time." To do this:

> "Complete data on every load in every boxcar, on every switch engine, on every train, in short, all detail data on everything involving rail operations, was fed into a master electronic Brain in each of the large cities served by the railroad. This Brain is connected to the local branch offices of all the other railroads, and it also gets information from all the other Brains in the United States and Canada."

Confusion occurs because an apparently minor detail is omitted from the program. Although the men working on the train involved have reasonable suggestions for the solution of the problem based on their experience,

[4] "A Hank of Wire," editorial, *Ithaca Journal,* August 31, 1961, p. 6.
[5] Anton Lee Baker, *Astounding Science Fiction,* August 1958.

none are feasible within the framework of the computer system, and the computer personnel have great difficulty in dealing with the situation.

The care needed in communicating with a computer is again brought out in "Spanner in the Works."[6] A person investigating the peculiar behavior of a computer is told:

> "...Naturally with any computer you've always got to remember the thing hasn't got eyes and ears, and if you don't tell it something, it doesn't know. And every now and then, even if you're a cybernetics expert, you phrase information or questions ambiguously or plain badly, and the computer gets you wrong."

This story also emphasizes another very important theme—the responsibility for decisions remains with man. Here a computer called Genius is used by a security agency called the Intelligence Department. The machine contains a vast quantity of information and is programed to solve security and espionage problems. Because of the secret nature of most of its information, it can communicate directives but not the data used in arriving at them. Under its first administrator "the Genius didn't have the final say. It was, after all, only a tool. ... When the logical decision wasn't the right one, it was up to the Security Chief to divine this and take appropriate action." He sometimes "authorized slower, less certain but more humane measures." However, after his retirement the next administrator relies on its decisions completely. Because these decisions are so often in error, the security chief commits suicide. His successor is faced with the problem of determining how and why the computer apparently changed sides in the cold war.

Another story where yielding responsibility for decisions to a computer results in chaos is "Cybernetic Scheduler."[7] The use of a computer for registration and scheduling in a college is extended to include pre-registration counseling. The students are told by the computer the subjects they are best suited for, and the faculty are told the courses they are best suited to teach. The president of the college is fired when 3,000 students are told to change their majors, 2,000 are told they are wasting their time and money, some faculty members are told to take instead of teach various courses, and other faculty members are told they should not teach at all!

*   *   *

The problem of utilizing computers in decision-making processes is of much interest to those involved with automation because the task cannot be reduced to a simple step-by-step procedure. The effort expended by computer specialists in programing computers to play such games as chess is usually part of an attempt to further understand decision processes in a changing situation. The problem was well explained by Edgar Allen Poe in 1835 in presenting arguments as to why an automated chess player had

---

6 J. T. McIntosh, *Analog Science Fact and Science Fiction*, March 1963, p. 67.
7 Edd Doerr, *Computers and Automation*, November 1958, p. 24.

to be fraudulent.[8] Although Poe is regarded as the first American science fiction writer, the article containing this argument is serious exposition:

> "From the first move in the game of chess no especial second move follows of necessity. . . . But in proportion to the progress made in a game of chess, is the uncertainty of each ensuing move. A few moves having been made, no step is certain. Different spectators of the game would advise different moves. All then is dependent upon the variable judgment of the players. Now even granting (what should not be granted) that the movements of the Automation Chess-Player were in themselves determinate, they would be necessarily interrupted and disarranged by the indeterminate will of his antagonists."

The idea that man is responsible for making decisions is seen in the science fiction stories involving computers programed to play games. In "God and the Machine"[9] a computer is programed to play draughts. The programmer prefers machines to people because "deliberate cheating and lying . . . are phenomena confined to the human species" while machines "had about them the original and unsullied truth and honesty that he found lacking in mankind." He, however, includes in the program the examination of a random digit if the computer is faced with losing. If the digit is even, the machine is instructed to cheat.

In "64-square Madhouse"[10] the necessity for human intervention again appears. A computer is entered in an international chess competition. The programmer is a psychologist with a good knowledge of chess, and he insists on the right to adjust the computer program between games. The adjustments are made on the basis of his knowledge of the individual characteristics and weaknesses of the next opponent. The story contains a clear description of the chess-playing program and indicates the usefulness of "a programmer-computer team, a man-made symbolic partnership."

*   *   *

One of the most persistent, and perhaps also the most important, themes in science fiction has to do with man's need to be creative. He cannot delegate creativity to machines. For example, in "The Feeling of Power"[11] computers have become so advanced and widespread that no one knows how they work. A man discovers how he can add, substract, multiply, and divide with the aid of only pencil and paper. The people find it hard to believe that a human being can do what a computer can do. The creator of "graphitics" commits suicide when he sees that his discovery is to be used for destructive military purposes instead of the benefit of mankind, but another person who has learned the process from him has a feeling of great

---

8 "Maelzel's Chess Player," *The Complete Tales and Poems of Edgar Allen Poe* (New York, Random House, 1938).
9 Nigel Balchin, in *Fantasia Mathematica,* edited by Clifton Fadiman (New York, Simon and Schuster, Inc., 1958).
10 Fritz Leiber, *If,* May 1962, p. 64.
11 Isaac Asimov in *The Mathematical Magpie,* edited by Clifton Fadiman (New York, Simon and Schuster, Inc., 1962), p. 3.

power and satisfaction from knowing, without the use of a computer, that $9 \times 7 = 63$.

Perhaps there is a moral here for businessmen who, absorbed with the rational, economic view of automation, find it hard to understand labor's lack of enthusiasm for new installations. Is there more to the problem than objections to job changes and temporary unemployment? Does resistance spring also from a lurking fear that computers will take away some of man's creative tasks?

The Silver Eggheads[12] has a related moral. This story centers around a search for writers after "wordmills" are destroyed during a revolt of authors who have become merely ornamentation. The wordmills, computers that produce fiction works, had been introduced toward the end of the twentieth century when publishers found that they were easier to handle than creative writers. The publishers had felt that "the two activities involved in writing are the workaday unconscious churning and the inspired direction," and that the former could be done by machine while only the latter, the "ulti-mate directive force," required a person—called a programmer rather than an editor. Gradually the public taste had conformed to the type of reading fare produced by the wordmills. After the machines are destroyed, writers are faced with the realization that they cannot arrange words on paper in any pattern and cannot "visualize starting a story except in terms of press-ing the Go Button of a wordmill."

Do the events in this story seem implausible in light of the following report of a project at Massachusetts Institute of Technology? Some of the reported goals of the project are to analyze works of art, to gain deeper in-sight into creative processes, and to have the computer perform mundane, time-consuming tasks. According to the reporter:

> "Original subplots for typical western TV dramas have been written.... As handled by the machine, this subplot is a sequence of phrases such as: the robber takes a drink; the sheriff fires a gun; the sheriff enters the room. All essential details of the characters and the setting are taken into account, as well as the logical relationships among them. For example, the program con-tains a provision for the robber's actions to become less and less rational as he continues to drink. Variations are obtained by making certain choices accord-ing to random numbers. Results when acted out seem to be on a par with scenes from class B western movies."[13]

The attempt to utilize computers for creative purposes succeeds far too well in "The Ultimate Copy."[14] An advertising agency buys a computer whose primary purpose "is to produce creative ideas that will stimulate the emotions and the intellect into a desired reaction." By combining data on the product (in this case a particular brand of cigarettes) and theories of psychology, the computer produces a "mathematically correct formula"

[12] Fritz Leiber (New York, Ballantine Books, Inc., 1961).
[13] Joseph A. Thie, "Computers in the Arts," *Computers and Automation,* September 1961, p. 24.
[14] Mel Carlson, *The Adcrafter,* May 1957.

which contains "all the intellectual and emotional stimuli necessary to produce the desired reaction," and from this writes copy for any specified media. The first copy produced by the computer causes sales of all other companies to drop to zero, but when a competing agency buys a similar machine, the public desire to obtain cigarettes becomes so great that consumers begin raiding and rioting.

Again we see hints of this fictional future in current events:

> • One report states that "there are many indications that the decision-and-action process of consumers will soon be simulated, or at least reduced to a systematic set of mathematical functions which can be programmed into an electronic computer."[15]
>
> • Another write-up suggests that by using linear programing and "considering the various motives for buying and not buying the optimum theme for the advertising may be developed."[16]

All of the cited themes which emphasize man's responsibility appear in "2066: Election Day."[17] By that time, we are given to understand, the process of election has changed drastically. Anyone who chooses to can take tests for the Presidency, and the computer, called "SAM," selects the qualified person. On election day in 2066 a problem arises: no provision for selection has been included to cover the event that no qualified individual has applied. A group of men who know precisely how the program operates are able to fool it into making a selection. Because the President's job has become still more difficult than in the times we know, the computer is also used to assist him in routine matters and in the making of decisions. The outgoing President, fearing that the President-elect will rely too heavily on the computer, cautions that:

> "A machine is not creative, neither is a book. Both are only the product of creative minds. Sure, SAM could hold the country together. But growth, man, there'd be no more growth! No new ideas, new solutions, change, progress, development!"

*   *   *

There is another side of the computer as seen in science fiction. This is the public relations aspect. The example I shall cite—"Alicia Marches on Washington"[18]—is noteworthy because it appeared in *The Saturday Evening Post* and so probably received the widest circulation of any of the stories that have been discussed.

This tale extends the current use of computers by the Internal Revenue Service to the point that "the boss" is a computer. A woman, trying to return $86.16 that was erroneously refunded to her, asks in hushed tones,

15 C. Joseph Clawson, "Simulation of Consumer's Decisions," *Computers and Automation,* March 1959, p. 12.
16 F. R. Dornheim, abstract in *Computing Reviews,* March 1961, p. 51.
17 Michael Shaara, *Astounding Science Fiction,* December 1956, p. 44.
18 Robert W. Wells, *The Saturday Evening Post,* January 21, 1961, p. 25.

"You mean they have the final word?" The reply: "Of course. No human brain could grasp the complexities of the system. These machines are the only ones capable of understanding what it's all about." Evidently even they are not, however. Despite such statements as "We do not make mistakes" and "To err is human," the woman cannot accomplish the return of the money.

The story's portrayal of a computer is well described by the *Post's* blurb on the author, which says: "He convinced himself that something more fiendish than a mere human was in charge of the system." This story suggests that impersonal treatment and almost impossible communication are associated with companies or agencies that use computers.

*   *   *

Now let us put specific titles aside and reflect in a more general way on the themes which run through science fiction.

The theme emphasizing man's responsibility for careful usage extends to all levels of association with computers. Most people who work directly with computers realize the trouble that can arise from an erroneous instruction in a computer program. However, the difficulties brought about by overlooking, misinterpreting, or underestimating various facets of a problem when it is being analyzed for computer handling can be more far-reaching.

Although computer specialists can be of great assistance, the responsibility for accurate and sufficiently detailed analysis of a problem or system also lies with those whose major concern is the problem itself. It takes a close and continuing collaboration between computer specialists and specialists in the problem area to bring about an analysis that recognizes both the intricacies of the problem and the intricacies of the computer. Analysis and collaboration can afford an opportunity for re-evaluating and gaining new insights into a problem that has appeared straightforward; it can bring to light assumptions or limitations that have been ignored.

Once a problem or system is apparently being handled properly by a computer, man's responsibility for creativity does not diminish. Despite the time and effort expended in the preparation of the computer program, new creative approaches still need to be sought. Modification can be achieved more easily if the fact that it need take place is kept in mind from the beginning. Ideally, the utilization of a computer should enable more human effort to be directed toward the aspects of the problem that require imagination, sensitivity, creativity.

Science fiction's emphasis on man's responsibility for making decisions is another theme that must be thoughtfully considered as computers are assigned to more and more tasks. In some applications the computer only analyzes the information presented, and the actual decision is left to a human being. But in many of these cases the final decision is merely a last obvious step following from the way the program has been written, the questions have been chosen and asked, and the criteria for germane information have been selected. In other cases there are frequently people who are responsible for approving the decisions before they are implemented. Does the use of computers speed the decision and reaction cycle to such a degree that this approval is not based on sufficient contemplation? That is a danger.

Computer applications have progressed from aiding in decisions about the manipulation of inanimate objects (e.g., financial data) into situations which involve the manipulation of people. Where these situations are expressed mathematically, it is questionable that the representations can really consider the pertinent human factors. As computers are applied to more policy decisions, the need for utilizing social and psychological theories increases. Although such theories are not as exact as those in the physical sciences, attempts are made to cast them into a logical and mathematical framework. For example, equations describing human values, human motives, human behavior, and even numerical values for human lives have been developed. While it is doubtful that these representations are adequate, the gravest concern is that decisions and activities based on them encourage a philosophy of expediency rather than humanism and lead to an environment where belief in mass society overrides the uniqueness of the individual.

Another issue which becomes apparent when reading science fiction stories and popular news articles is the public image of computers. The moral of the literature is easily drawn, it seems to me. Those who are even remotely associated with a computer installation should avail themselves of the opportunity to understand the processes, capabilities, and limitations of computers, not just to see other possible applications in the same setting but also to gain a realistic appreciation of their present and future roles in other areas of our culture. The information should then be brought to the public's attention, as people are evidently in need of a greater understanding of how automation stands to affect their interests.

THE COMPUTER IMPACT
Edited by Irene Taviss

Here is an exciting new book that examines the social implications of computers by looking at their impact on the economy, the polity, and the culture. Following a brief discussion of computer methods and uses the book turns to:

- computer application in industry and the services, management decision-making, white collar workers, and — more futuristically — the computer utility, the credit card society, and the leisure society;

- computer applications in government, political decision-making, social planning, foreign affairs, the privacy question, and the law;

- computer applications in the schools, thought processes, the sciences, social sciences, the arts, libraries and information, and attitudes and philosophy.

A general introduction and four introductory sections place computers within the context of contemporary social patterns and thus help to show the reader the ways in which technology both shapes and is shaped by the society in which it is developed.